T0070282

THE COMPLETE STUDENT DESIGN PRESENTATION SOURCEBOOK FOR THE PRACTICE OF ARCHITECTURAL ENGINEERING

ATA ASHEGHI

THE COMPLETE STUDENT DESIGN PRESENTATION SOURCEBOOK FOR THE PRACTICE OF ARCHITECTURAL ENGINEERING

Copyright © 2022 Ata Asheghi.

All rights reserved. No part of this book may be used or reproduced by any means, graphic, electronic, or mechanical, including photocopying, recording, taping or by any information storage retrieval system without the written permission of the author except in the case of brief quotations embodied in critical articles and reviews.

iUniverse books may be ordered through booksellers or by contacting:

iUniverse
1663 Liberty Drive
Bloomington, IN 47403
www.iuniverse.com
844-349-9409

Because of the dynamic nature of the Internet, any web addresses or links contained in this book may have changed since publication and may no longer be valid. The views expressed in this work are solely those of the author and do not necessarily reflect the views of the publisher, and the publisher hereby disclaims any responsibility for them.

Any people depicted in stock imagery provided by Getty Images are models, and such images are being used for illustrative purposes only. Certain stock imagery © Getty Images.

ISBN: 978-1-6632-3333-2 (sc)
ISBN: 978-1-6632-3332-5 (e)

Library of Congress Control Number: 2021925212

Print information available on the last page.

iUniverse rev. date: 12/14/2021

The Complete Student Design Presentation Sourcebook for the Practice of Architectural Engineering

A complete sourcebook for how to create architectural and building design presentations. With design strategy chapters using the right media for applications in architectural engineering design. Intended for beginner, intermediate, and advanced studies in architectural engineering.

A great sourcebook for students, professors, and professionals.

With Images and Diagrams

Ata Asheghi

I dedicate this sourcebook to the young students who wish to pursue the fields of architecture, engineering, and design construction. I made a goal to assist those who may have difficulty in this environment and who face the adversities I have faced throughout my schooling up to this day. The fields demand more than you expect, and every detail is judged, so the purpose of this book is to prepare those who need to know what to do and exactly how to do it. These are words that you can be encouraged by to enhance your ideas of how to create the solutions you need in such a diverse and unconventional career life. I further dedicate this sourcebook to the admirable university professors around the world, from all walks of life, with the wish to enjoy the spirit of design and creativity. There is a silent admiration for learning and reading that gives us readers so much insight into the world around us, with the expectation of knowing where things lead if they lead anywhere. We also know that the experience held in viewing, visiting, and living in places gives us certain conditional aspects of place, time, and presence. Trying to step out of a box is not easy. If that is the goal, then as the reader, you must understand that the box was not built for you to be kept in so you can ignore what is outside, nor is it for hiding what you know in. There is nothing keeping you from opening it. For those in life who have expected more for themselves and tried their best to make some sense of change in a world where everything changes, I give you credit, and I give you this sourcebook. Nothing and no one can take your talent away from you. I am living proof.

—A.A.

L'excuse n'est jamais la réponse, mais la réponse n'est que la question.
(The excuse is never the answer, but the answer is only the question.)

-Ata Asheghi—

Endurance is not what you have to absorb in order to learn.

-Ata Asheghi—

CONTENTS

INTRODUCTION

Each chapter in this technical sourcebook is to be used to lead the student into conventional and innovative methods of applying conceptual design to both 2-D and 3-D applications. This is meant to strengthen the learning abilities of the avid designer and help maintain the traditional applications of the design architectural world through fundamental drawing techniques. With a set of six strategy chapters, the student will learn how to arrange and organize designs and tailor them into a sequence of architectural, construction, engineering, and design drawings by incorporating work into two-dimensional (2-D) and three-dimensional (3-D) environments. These techniques allow work to be visible to the reviewer and will give the student the ability to manage workable ideas using planning and organizing drawing methods. The main aim of this book is to teach the student to rely on his or her constructive drawing techniques with the use of the design examples provided. This will offer a gradual learning experience with respect to the student or professional's criteria for improving techniques. Concept design, drawings, sketches, and drafting methods are incorporated into several guidelines to provide the essentials of drafting and drawing used by beginner, intermediate, and advanced designers.

The sourcebook is based on the collective academic teachings that I, as the author, gained through architectural design courses. These teachings are meant to reintegrate a sense of practice into the basics of pencil and paper. In terms of the basics, we will explore the importance of what was and still is today a means of discovering creativity while tying in a new-age evolution of computer-aided design. To accommodate the student with the right tools, the following objectives can aid the functional planning of design drawing:

1) The "how-to-use" designer drawing materials
2) Sketch drawings, concept sketches, and overlay trace sketch drawings
3) Free-hand drawing, line work methods, scale, and proportioning
4) Pencil and pen drawing, pencil and trace overlay, and pencil and pen drafting
5) Drafting techniques, drafting organization, and drafting sheets
6) Model-making and preparation/fabrication
7) Program development, concept research, and materiality

8) Digital computer aids (Revit Architecture) drawing transition/translation
9) Architectural and construction methods application
10) Final presentation preparation and organization

This sourcebook is written to guide the student/designer through a design process by incorporating hand-drawing techniques and ultimately applying them into computer design software, providing a detailed means to produce working architectural and construction drawings. This transition/translation method will be explained and will also enable reviewers to view the interpretational methods used in both medias. The process is detailed and will provide a suitable alternative to creating office workflows in the forms of design sheets and design drafting sheets while enabling a compatibility with the appropriate transition into the computer software. The designer will be more aware of his or her own designs as he or she learns effective self-editing techniques throughout the organizational process. Directives on how to produce, organize, and execute specific drawing applications will be examined and will allow the student/designer to understand the ways in which the right tools are essential in completing design presentations, which is the overall objective of this sourcebook. The handbook takes a look into the history of drawing and drafting and discusses a model approach to defining the aspirations of the designer. A conversational chapter is also included at the end of the sourcebook to emphasize the tactile techniques that many beginner designers implement in their design methods and help delineate the outcomes for the avid architectural engineering student. By outlining the traditional evaluative design strategies used by predecessor designers, the new-age designer will learn to be mindful of the replicative application of past and present design methods.

For the student or designer involved, it would be beneficial to learn the practices explored within this sourcebook. This handbook would also ultimately be of aid in the curriculum of architectural school. By knowing these design strategies, a student may excel in design classrooms and may also enhance his or her own personal design development. The student will learn to become a designer by achieving these practices into routine and then later into semiprofessional or professional careers. It is possible for the student to quickly learn these techniques and, in doing so, develop a greater aptitude with the knowledge of these applications.

PART I

FUNDAMENTALS IN ARCHITECTURAL ENGINEERING, HISTORY AND ROLE

1

THE DESIGN WORLD

The design world can be explored through interpretation and reinterpretation of ideas integrated by means of applications of various mediums. These mediums have been refined over many years and have given the current outlook on today's practices a new perspective. Every visual sense of the world around us has the implications of design applied into it, whether it be buildings, artwork, graphics, or even daily household items. The sense of design illuminates our creative ambitions and shows a greater constructive aspect of ourselves as we change the physicality of the world around us into what can better suit our needs. Design incorporates the use of line, form, color, shape, quality, distance, light, darkness, durability, clarity, and many other aspects that render it a composite of all substances. For the designer, being able to address each component of design greatly impacts the overall design use approach to whatever design project is being made.

In the fields of architecture and engineering, the role of design is essential in the formulative program-oriented outcomes that building programs require, and the development of architectural systems. It is linkable to the primary conceptual nature of drawing in that the fundamental systems involved to produce concepts and drawings are the first steps toward the use of design. Design preconditions where the student has observed the programmatic aspects of a project are generated through the initial phases with sketching and drawing as a foundation. This serves as a preliminary idea configuration for the designer, and it would most certainly be the stepping-stones for what would come next.

Enhancing the primary abilities of drawing on a 2-D format by utilizing proper technique and adjustments to line work is essential in this practice. Drafting is one aspect that requires special skill in the discipline of finite hand-to-eye coordination, which effectively seeks to mitigate error in design production. Otherwise, the initial design traces can be corrected with pencil and then later redrawn or marked over by pen. This is the basic conventional approach to standard drafting practices; however, there are specified techniques that are crucial to the overall effectiveness of each use of line work, line weight, drafting-drawing revisions, and several other incorporated practices to produce quality work. There are a variety of factors that contribute to the way in which drawings are produced whereby is a culmination of factors that can be set to how drawings are produced, development steps are proponents to the procurement of skill and accuracy in implementing them in a project. Most of these will be examined in the sourcebook. On the how-tos for each design strategy chapter, diagrams and representational concepts will be explored to show how to utilize the skill of line making.

Lastly, this sourcebook is for anyone who wishes to obtain the skills needed to approach their education in the fields of architecture, engineering, drafting, technology, and design. This book can be used as a tool and reference for the initial and intermediate design phases of a project, whether it be for the design of a building (in this case specifically the fields of architecture and building design), a product, or a graphic visual or design for the sake of refining artistic abilities. The book *does not* limit the techniques that can be self-accommodating to the student while reading, and it encourages students to self-explore potential in their own design methods. The principles of hand-drawing techniques are explained in detail with the emphasis on the individual replicating, reproducing, and innovating his or her own talent.

In sync with the fundamental practices of the early stages of hand drawing, a transition into computer modeling is also important to this process. Computer-aided drafting and modeling software reintroduces the same techniques used by hand and of course with greater versatility into today's digital direction. Looking at these software programs gives us a better advantage in developing more accurate sets of working drawings. Although a few of the software applications may be difficult to learn on their own, this sourcebook will direct students in their basic uses from the 2-D environment, which will be discussed. This is the step where learning the software is secondary to designing the drawings. Computer-aided design software programs such as Autodesk Revit Architecture will be outlined, and specific design-oriented command functions in these programs explained to give the necessary tool-action settings to create three-dimensional (3-D) modeling and computer-drafted line work. The applicable methods to incorporate 2-D hand-drawn work into the three-dimensional or two-dimensional digital format will show the reader the depth of this greatly suitable means to communicate construction language. By the implementation of computer-aided software packages, the dual coordination of drawings to computer software becomes the new tactile method of design incorporation.

Because of commercialization and legal implications of the software mentioned, this sourcebook will provide obtained instructions from a text featuring the command functions of the software in a definition-style section. To adhere to the sourcebook's intent of allowing hand-drawn designs interpreted into computer modeling, it is necessary to note that these instructions will follow with the licensed intellectual property produced by the creators of the program through sourced information from their textbooks. Academically speaking, this in a sense is a reinterpretation of an Autodesk software guidebook in an instructive and step-by-step fashion. The sourcebook is solely intended to guide the student through coursework in college-level design courses that teach commercial- or residential-style architectural design. This was the determining approach to the outlining idea of the sourcebook.

To allow the designer or student to comprehend the variable aspects of both tactile and digital formats, the last chapter makes the student aware of the importance of this for occurrences where editing and revision are completed before the submission of drawings into the digital environment. Reducing the frequency of mistakes, allowing flexibility to design in both applications, and making the design process enjoyable are the goals at the end of the chapter. And, as mentioned, a conversational walkthrough will be included to address questions one might have about this method in detail. As the reader, please take the time to conceptualize your ideas and preparations for your future work!

Design is a plan for arranging elements in such a way as
to better accomplish a particular purpose.

—Charles Eames,
Architect and Designer during the International Style of Architecture

HISTORY OF THE DESIGN METHOD

2

Dating back to the early Renaissance in Western Europe, midcentury transitions were made with the advent of printing, drawing reproductions, and the fine arts. During this time, art and drawing on paper became a more advantageous tool with the arrival of new yet stylistically simple drawing tools. The sense of maintenance of the drawing during these times became a response among many artists, who wanted to enhance the skills used to render more vivid and accurate works. Artists began utilizing a diverse pallet of techniques with different mediums to surround their skillfulness in drawing with precision. The drawings of the early renaissance and later styles showed the era an assimilation of old technique with new technique and vice versa to a point where drawing and painting became parallel to real-life imagery. The profound works of Leonardo da Vinci, Francisco de Goya, Monet, and Rembrandt had shown the ages the immeasurable skill of drawing production in which the artistic capacity to draw into realism transpired congruently with the interpretation of ideas through detail drawings with artistic styles.

Surrealism, realism, and impressionism later developed from the pinpoint senses of modernity through more refined pencil techniques. Work produced during the times homed in on clarity and definition while at the same time inspiring other artists to commandeer monumental paintings. They drew inspiration from many of these artists, and uniqueness was born. Famous works of art featuring it still exist. Inherent in the works of the many artists and thinkers of the time was the central underlying notion that specific times called for both prerequisites in design themes and new discoveries of the use of modern design techniques.

The ways of design used during the mid-seventeenth, later eighteenth, nineteenth, and twentieth centuries had evolved through the exploration of innovative technologies and approaches. As earlier stages of design were conceived throughout the midcentury, so did the development of the term *design*, being the result of a plan, specification, or procedure used in the form of the construction of an object or system. Slowly maturing into an observational practice of complex yet corresponding linkable arrangements, design had been improving on its way toward modern versatility. The centuries-old traditional methods of drawing line work and sketching became more and more technical into the advancements of detailed line work implementation, diagrammatic sketch work improvements, and the geometric advent of the "the grid." The sketch had evolved into more than a means to show imagery, as later developed in the late eighteenth and early nineteenth centuries; what composite sketches had produced during the time became the conventional standard.

Drafting and free-hand drawing techniques enhanced the versatility of the drawing pallet, leaving a marker on the industrial society. The designer became synonymous with advancement in technique, and throughout the years, improvements initiated more practical, "idealist" characteristics in the designer.

Design led the way toward the new modernity of the industrial age and the rise of the first design-developed steel-framed building. In its proprietary sense, the use of design had been prevalent in reshaping the perception of what was slowly becoming obsolete into newer types of innovation. In this century's design-oriented world, continuous improvements recategorize and objectify what is new from what was old. This sense has been a universal trend or trait marker for the designer. For example, the Greek Acropolis, having been designed with many considerations to both architectural and mathematical dimensions, showed comparative influences from other nations of the time. The adherences to new methods of construction, form, and design of the Acropolis brought into focus the very same improvement advancements made to its predecessors. This response carried the language of asking what the different approaches to design were compared to later design engineering developments. The modern nineteenth-century era, with the steel-producing backbone of the times, gave way to entrepreneurial talents who shaped landscapes, erected tall buildings, and conceptually grasped the array of design philosophies that incorporated various methodologies at work.

For Gustave Eiffel, at the time a world-renowned designer-architect, design was central to the task of completing the Eiffel Tower in Paris, France. We see the tower today as a testimony to the free-willed spirit of the designer during the time, where technology and adaptable circumstances reflected the intention of design comparable to the centuries-old systematic and untethered approach to drawing. However, the era of the great Roman Empire, along with those of several other empires of the ages, advanced the engineering discipline in systems used to create palazzos, bath houses, aqueducts, roadways, and the like, which existed through influential design. Can we see commonalities in both the past and present with the continuous refinement of design?

We turn our own pages in society. Having done so more fervently over the course of the past two decades, we are now in a new decade of dominance, a decade of overlapping, reused, and articulated design-integrated surroundings stemming from a reemergence of traditional expectations. As it is said with historical eloquence, it "has been done before." We are most certainly at a new stepping-stone in terms of the design world. The many generations who lived and saw change in the world, whether it be from city to city, town to town, rural area to suburban, or coast to coast, were forever centralized on the advent of commercialization, which is now leading humanity's new change. The role of design has become in *our* modern times a new fever in that, depending on who chooses to multiply the dream, the sole purpose of it is to improve responses to previous conditions, which were themselves responses to other previous conditions, ultimately becoming a syndrome of repetitious perfection. To repeat! We make the same mark again and again in order to clarify an idea that was gained, lost, or renewed.

Design, or whatever is the so defined term, can lead to the same result that was conceived initially through trial and error. Lastly, the works of the many famous architectural engineers and architects of our present time, including the many star architects who, over the last few centuries, have built stylistic and categorically profound architectural movements, have certainly become our new influences. New symbolic representations of what was, what has been, and what is enlightened our curiosity. It may seem as if the shadow that has been cast from these buildings connects with our lives and the lives of those who built them in a way that I personally refer to as a "shelter of ideas," in that people of our time can facially recognize these places and objects at their own discretion with the main concern for the past again being led into what is new-age territory. But even spiritually speaking, our age of times, places, themes, and buildings is testimony to our stance here in this century. It is also a tragic view that this is part of humanity and part of the design world we live in, in which everything old becomes new, and everything new becomes old.

AUTHOR'S ROLE AND STATEMENT

3

Some of the greatest and most influential people of the times have disassociated themselves with academia and in respects have pursued their talents beyond the traditional norms of conventional societal institutions.

-Ata Asheghi—

I want to be able to introduce ease into the process of developing documentation and help you, as a reader and explorer, implement a good work ethic to produce what would be called a change. This sourcebook will guide you through your design intentions if you are a student newly participating in architectural engineering studies, you are a researcher trying to develop effective solutions for your preliminary projects, or you are a perfectionist in aiming to develop your ideas further with your sense of purpose. The information provided is derived from the continuous application of architectural drawing methods I've researched throughout the course of my (the author's) education in conjunction with academic research implemented intermittently throughout my schooling and beyond. Academic skills that I gained through design courses I took during my college career have all included the hand-drawing approach to designing architectural projects on large two-dimensional surfaces, enabling a tactile sense of learning and creativity to prosper.

The work within this book is somewhat of a condensed version of the program I had completed in my undergraduate classes studying architectural engineering; however, I do not replicate the teachings and academic research conclusions from the years of my study in this sourcebook. Rather, I offer this to reproduce the same knowledge. This would not be considered a replication of existing drafting theories, per se, but casually addressing it, it would be appropriate to see that I, the author, note that the techniques and skills addressed here are formulative rather than academically taught.

My take on drawing is that it is a creative talent that can be refined. And through refinement, one can produce a greater opportunity of applying a composite of knowledge.

As a personal note, as I have openly expressed, I believe that in writing this sourcebook my motivation for the reader can be grasped through the sense of an idea of applying what is conceptually produced within the learning brain to what may be the physical environment. Through my studies, I incorporated problem-solving techniques to define solutions for work-related projects in renovation, home improvement upgrades, assistance with client designs needing visual planning, and personal/theoretical portfolio projects that refine the skills of practical construction methods. Most of the information taught to me by esteemed professors and skilled professionals was what I term a "click skill," as the skills developed and addressed in studios repeated the applications, which clicked with each skill necessary to create projects. One example would be drafting on the drafting board and drafting in pencil and then later revising and editing in pencil and finalizing in pen.

In my writing of this sourcebook, I wish to reiterate the notion that design is an indispensable tool used to implement basic art principles into dynamic and versatile techniques toward the creation of a project. My education in architecture and engineering is the foundation for this book in every sense of each chapter. I personally have had the goal to write this in effort to educate beginning designers and learners in these fields of study to make them aware of the solutions for using the right methods to create quality work. I also felt that this would be a way to minimize overusing unconventional strategies and to incorporate strong visual methods that can transpire into the language of technology. We the designers and career-minded professionals in the fields of construction and architecture aim to promote the versatile resource of the design language through what we present visually and buffer the idea of identifying programmatic issues that encompass each design perspective through means of refined interpretation.

The chapters included show the finite practices of architectural drawing, drafting, and organizational planning along with theoretical elaborations related to the fields. To emphasize more deliberately, the sourcebook can be used not only as a side reference book but also as a documented resource journal, which may serve to aid in the discovery of new theories or new conventional and unconventional methods of design. Included in the book are several empty note pages for the reader to generate his or her own ideas or strategies. With the use of these chapters and design resources, we can then see for ourselves the creative capacity of what has been most often seen in the exterior environment as "planned ideas."

The construction industry from the early beginnings to its advanced and rapid growth during the Industrial Revolution has seen a multiplication of drafters, designers, and layout and drawing technicians, all whom have mastered the fundamental principles of hand drafting. Drafting techniques were at the time thought of as highly valuable, and with the advent of drafting equipment tailored to reinforce the techniques used to articulate drawing line weights and line work, drafting technicians became more and more versatile. In this sourcebook, I approach what is or has been conventionally taught to professionals as standard drafting techniques and apply them to the new-age three-dimensional computer modeling agent, which is now the industry standard. Obeying this strategy not only helps produce greater sets of drawings and offers a greater capacity to invent, revise, and relay drawings but also stamps on the idea that work is produced. The work, and the methodology behind this, is subsequently an elementary form of basic hand drafting, which is interfaced with the dual application of the computer. Knowing that the drafting language or line-work language is somewhat set in stone in the hand-drawn world, in this sourcebook, I will guide the designer/reader/student/ professional through steps on how to showcase, produce, and refine drawings to be reused in 3-D modeling software. The virtue of this is that work and time spent on projects can have potential monetary value when considering an employer's or employee's agenda. This not only reiterates how work is spent producing fundamental drawings but also abbreviates the form of design strategy used with a purpose that could no less be done by the computer itself, however without the limit of preliminary design.

To put this idea into words more reasonably, again, the preliminary design intent is the use of hand drafting and the observational, new-age use of drafting software, which would be in sync to finalize a completion method. With the theory now set aside, we will explore in each chapter the sets of strategies to help give you the skills you need to complete your projects and a creative outlet for you to dispense your talents and passion into the world of design, where you can make a difference in what is around you.

As you read each chapter and combine the techniques presented, you will learn to see each new idea with intrigue. It is the art form that brought you to your desire to draw, and it is hopefully the dollar that has shown you that you can take your ideas to the bank.

4

DRAWING DESIGN TOOLS, METHODOLOGY, AND FUNDAMENTALS

Drawing and design hold a great value in the foundation of artistic abilities that may be performed, from simple sketches to large, full-scale plans. These are very useful talents, and they subjectively tell us that creativity is a reward. Drawing has its fundamentals in pencil and paper sketching, allowing for the activity of ideas to respond to a 2-D format. This enables the individual drawing to interpret conceptual ideas or schematics onto a physical environment. With paper as the means to develop visibly rendered ideas, other tools and applications provide interpretation formats for manifesting conceptual form. The materials selection in the next chapter helps to address the types of studio-related tools needed to perform various tasks in drawing, drafting, and color-producing images. These materials are hand selected by the author, and images have been provided along with good descriptions to define the tool and the task.

This chapter serves as a written walkthrough for the specifics on the methodology behind and use of drawing and art tools, which are important to use during early and intermediate stages of studio-related drawing production. The tools and procedures associated with them are integral to the application of strategies and to producing quality works. Along with the corresponding images to each drawing application in the next chapter, the author's suggestions will be provided to aid in the process of the application of the best materials.

The materials listed can be found in local arts and crafts stores and are available for purchase online. The following are the materials used in design drawing, design drafting, conventional drafting, model-making, and studio-related design. These do not encompass the vast array of materials that can be found at your local or nearby arts and crafts store but are chosen for their importance in their studio-related use. The materials listed here, as well as their accommodating descriptions and procedures for use, are all meant to procure the right methods in later chapters.

This materials section is a step toward knowing the right office drawing and drafting tools to fit the bill for the best approach with consistency in design technique. Most items are found to be retail-priced and affordable, though some are specialty items that may only be found at specialty art stores. What is unique about these items is that they each have a certain versatility to them and could be used conventionally and unconventionally in terms of drafting design. By the term *conventionally*, I state that previous if not well-taught methods of drafting were standardized, but in the case of today's new tools and technologies, we must say in this book that an unconventionality or "advent approach" is being presented. This, in terms of refinement into the practice, can be adhered to as a new-age category of design use. Many of the world's greatest buildings, places, and structures have all in some way undergone some sort of preliminary or sketch-based idea, involving the use of drawing formats, sheets, and grids that can offer the designer an infinite pallet of ideas. Invigorating a sense of versatility is the main goal in this sourcebook, and to the liking of the reader, the drawing tools necessary for the execution of specific schematic and interlinked drawings are greatly sought after.

What we can see as necessary to the implementation strategies used to create accurate and acceptable drawings is a *precondition* that predecessor draftsmen and designers have done it before, meaning that their exploration of resource and quality use of line was the tenant of their practice. It may simply be put that these principles can be restated in another way to adhere to the knowledge behind new ways to procure refined drawings.

In our schools in the United States, through specific regional programs tailored to the fundamental aspects of hand-drawing, conventional methods on how to teach students how to emphasize their drawing-drafting skills are still offered to undergraduate students under this background. These schools are, in my opinion, the best option to teach students the methods used in today's modern culture. As a student of a private university in the early 2000s with a program still under development and curriculum being refined, I found basic drawing was the key ordering factor in the program's curriculum objectives. For many schools, design starts with the ability to draw, and the clarity of this idea is reshaping into what previous years have brought into the picture as the progressive application of modern computer-aided drafting software.

Many of the courses tied to design-drawing are linked to the diverse breadth of versatility of the hand-drawing pallet of design use, which may be the best start for students entering these schools.

However, there is a new custom in today's industry settings, along with modifications to architectural grammar, architectural configuring, and architectural modeling display communication, or (AMDC), which I term as being an off-breed facet of architecture that is leading the way in fields of building information modeling (BIM) and data management. For one, with the vibrant use of programs that undergo annual upgrades and package software platforms like AutoCAD LT Architecture, one of Autodesk's forefront software offerings, and Autodesk Revit Architecture, the new platforms engage most of what the previous CAD software hadn't done before, by allowing the best user interface and modeling techniques to reach maximum possibilities with the variety of drawing sets, construction plans, and schematic designs required for academic credit. Give or take, these astoundingly compatible software programs reduce the actions performed on the working drawing drafting desk, which eliminates all or potentially most common sources of human error.

Autodesk Revit Architecture, being by far the leading architectural modeling software, comprises line work use in the form of commands bulleted on the screen to be easily selected for various building operations. Along with making use of the many built-in resources to facilitate graphics commands, structure treatment, and composite rendering, the user can operate in the interface by learning the set of commands that delineate specific operations used to ultimately construct 3-D renderings of composite buildings. The set parameters in Revit are also analogously viewed as dynamic tool sets to accurately gauge and reset building design conditions that initially had substantial difficulty being produced in previous software. AutoCAD LT Lite and other versions operate on a similar platform and have different value parameters for their command use—these being the included functions features for drafting performed on the 2-D scale, for example highly recognized AutoCAD software includes a gridded background drop for line work use, which is the main interpretive command for the program, using specific annotations and functions in line work.

While Revit, being a modeling instrument, uses dimensional objects in the place of line-making to form structure or types of built form, AutoCAD emphasizes direct line creation. Both programs are extremely versatile, and in this sourcebook, while keeping direct reference to each program, several fundamental sourced software drafting techniques will be outlined. For the sake of distinguishing the two software programs for you, the reader, representational diagrams and images will accompany descriptions provided to guide the approach of hand drawing into computer application. The aim to utilize hand drawing and computer software application thrives on the techniques of visual compatibility, which will be presented to aid in the discovery of left-hand, right-hand work, or the coordination of tools.

MATERIALS

5

The materials we will use or see will directly depend on the drafting objectives we must acknowledge to produce quality work. The list of materials presented here is not all encompassing, but these are most likely to be the essential tools needed to preform architectural drafting. For the start of the list, we will go through what is intended for use by office professionals and what each of these tools or drawing tools is used for. Let us now look at what we need to start off in our design practice. You can purchase these at your nearest art supplies store, as they will have a large quantity of stock in items that are intended for architectural drafting and drawing while also having items that are for the sole purpose of drawing that may be compatible with the standards of drafting shown in this chapter.

To reiterate, the ability to improvise methods and techniques is universal to itself in this chapter, and the means to understand how each tool is used is explorative to the designer. Keep in mind that the prices of some of these products might persuade you to look for alternatives, as well as give you ideas on how to reproduce your own tools.

1. Drafting table

This is a conventional custom drafting table used for large-scale drawings, and the other (right) is a commercial drafting table. Courtesy of Ata Asheghi.

The drafting table is where drawings happen. A drafting table is a large, flat surface, usually made of a wood composite or in many cases a plastic or solid polyethylene composite. Regardless of which table you wish to purchase, they should all function the same as the older drafting tables. During the roaring twenties and thirties of the Industrial Revolution, commercialized drafting tables were large and made of solid wood. They were stuck together through grooves and adhesives, coated with clear paint, and left flat for paper to smoothly remain on the board. Today's tables are plastic, steel, and composites. The table can incline upward, using side arms that can remain in several positions, raising the table to different angles. There are also drawers on some of these tables for storing tools. Again, the products of today or even ten or twenty years ago have been upgraded and even used in a portable capacity.

The image shows a handmade drafting table using a large four-by-six-foot piece of composite wood material, surface coated with white paint, with hand rests and tiles that have been glued to a large piece of long board on the edge. The other on the right is what is commonly known as a commercial or industrial drafting table.

The drafting chair also in the photo behind the desk is specifically for the drafting table. It can be increased in height up and down by a lever on the side and is sleeker and smaller than an office chair to enhance maneuverability and turning radius. Most chairs for drafting can be raised to two to three feet or more above the ground for the table, as some tables are set high for drawing and standing.

2. Pencils

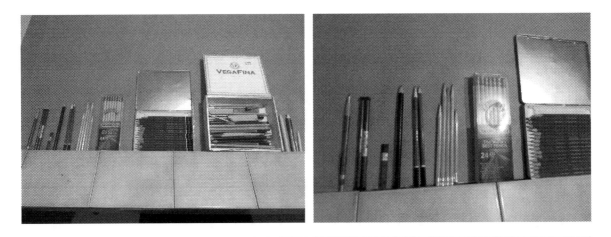

These are few of the many types of graphite pencils that are used in drafting, from 2B leads to H leads and so on, including the number 2 pencil. Courtesy of Ata Asheghi.

Pencils are common to students studying drawing or even doing notetaking. Of course, there is the notion that the pencil must be sharpened often to keep a fine point and have an eraser at the end to correct mistakes. But there are a lot of types of pencils, and all can be used to develop architectural drawings. For the number 2 pencil, the number signifies that the graphite core has a specific shade of gray that when applied designates a degree of blackness. The number associates with the hardness of the core in ranges of one to ten or even twelve—the higher the number, the higher the degree of shade that would be produced from the hardest core. The letters H and B associate with "H" being a hard core and "B" being a blackness or black applied core.

A number 2 pencil is an HB pencil that has both a hard core and a black applied core. These numbers and letters correspond to both the hardness and line weight of the core or graphite stem in terms of when the pencil is applied to a surface. The resulting line, depending on how much pressure is applied to the pencil, will give a total degree of darkness or scale to the white surface. Some pencils are noted F for fine point, as they have a particular sharpness that gives a quality of a darker shade of gray. Finding the right pencil depends on many factors, including the type of drawing you are doing, as opposed to the necessary line weights needed to draft solid lines versus lighter lines versus visible lines and underlines. The techniques for the use of these pencils are integrated in this book, and one can master a versatile drafting drawing document through the correct use of line eight through these pencils.

Here are some pencils that are common in art retail stores, including the number 2 pencil. There is a set of pencils to the right. These have the specific line weights associated with them. Some come in a complete set of line weights, while others may include pairs of the same. There are also mechanical pencils that operate with small slivers of graphite that are meant to be placed in the receiving end of a mechanical pencil. These are usually 2 or 2B type leads and come in different gauges, such as a 0.7- or 0.8-mm graphite lead.

To note, for drawing and sketching, pencils are the first tools to show graphic qualities. Some pencils are large and have larger and thicker cores and fatter wood. These can be used like a marker, and the applied core gives a thicker diameter of graphite shade. The most particular type of pencil in the drafting world is the automatic mechanical drafting pencil. It is like a mechanical pencil but holds a large sliver of 2B, HB, or HBB leads. These are not produced as often as they were before the days of hand drafting, but they are used specifically for drafting line weights and architectural substitution of number 2 or H pencils.

3. Pencil tools, erasers, sharpeners, colored pencils, chalk, and charcoal

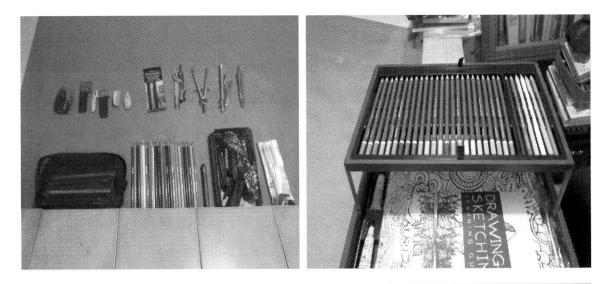

These images present a selection of drawing materials including erasers, colored pencils, chalk, protractors, and containers. Courtesy of Ata Asheghi.

For designers, possessing the essential knowledge of how to use materials is important to completing revisions and applying new methods. By this, a determination can be made as to which materials are needed to allow for specific changes in the drawing design process.

Well, for this selection, we are first looking at pencil tools and erasers specifically for the revision process and for adjustments through drafting techniques. These tools are what encompass the ability and the applicative technique that follow how the hands work in conjunction with the pencil.

The protractor is a steel or plastic tool that is shaped to the hand. Two pieces are met by a single divider at one end, allowing for the pieces to separate via an adjustable thread. This allows the two pieces to be distant from each other. Each of the tips of the pieces are fine points for grabbing onto a paper surface. The other type of protractor is one that may have a pencil fitted to one piece to allow for drawing. The drawing process is done by expanding or contracting the thread to allow for a specific distance between the pieces and in turn offer a diameter or radius around both points. This is done when one point is held on a surface, and the other point, whether it be just another piece or a pencil, creates a circle when rotated clockwise or counterclockwise.

The protractor offers a district variety of uses, and it is a historical drawing tool. It offers the ability to draw complete 360-degree circles without line error. It also allows for the measuring of distances by the expansion or contraction of the thread.

On most site drawings, plans, maps, or distance-related diagrams, there is a noted scale at the bottom. This may be interpreted to have been measured by some geographical measurement determined by the location or the distances between things. On maps, there are scales that represent the distances between countries and oceans based on latitudinal and longitudinal degrees.

The revision process in drafting and drawing is the immediate function to allow for changes on the two-dimensional surface. When lines are made and mistakes have been drawn, the solution to replace or clear those mistakes lies in use of the eraser. The eraser is a tool used to clear graphite drawn lines, and it is indispensable when it must be used for this purpose. The line work calls for accuracy, precision, quality, and completeness. When there are mistaken lines made, for the sake of the fundamentals of drawing, the eraser serves to edit those lines. By explaining that the editing process occurs when lines are taken away and drawn again, we see that the stimulus for change is made through the activity of the revision process.

Erasers come in different types, some small and some large. They are usually made from a natural or synthetic rubber that causes friction on a surface when applied and a reduction of the material when the surface contact is met with the friction. Some erasers, such as the gum-type, are stretchy to allow for the index finger to ensure accuracy in erasing in detail drawings. Some pens also may be erased through an ink eraser.

Pencil sharpeners are extremely important in terms of what they do for the pencil and for the designer. The sharpener creates a fine point to the pencil and gives the fine line work some accuracy when drawn. Sharpeners can range in diameter depending on the types of pencils used, and some are specifically for certain drafting pencils. The sharpener associated with the drafting pencil is an object that once applied to the graphite itself can sharpen it.

Drafting is a set of rules that coincides with how the pencil relates to the paper. For years, drafting was the equivalent of how people use programs on today's modern computers. Drawing requires fluidity, expression, and imagination.

For drawing, in terms of producing works with a pencil that need color, the best alternative to fill in line work would be the colored pencil. This is an experience worth attempting to show color in drawings.

Designers and architects throughout the years best displayed their work before the times of digital devices through the means of drawing, replication, and the interpretation of reality through color and line. The expressions made in these drawings capture buildings, places, people, and forms of life that have been rendered in a palette of colors. The colored pencil uses the same skill as the normal pencil but with the replacement of an inner colored core. All colors on the color spectrum have colored pencils associated with them. From rose red to aqua blue, colored pencils are versatile in applying colors to the drawing. They can also be revised through the eraser if the eraser is capable of clearing the color and the line.

Lastly, the colored pencil may also be used as an identifying tool, to create bold or lightly shaded lines for graphic line work use. These lines may interpret designations or highlights on the drafting paper associated with black lines, and they are usually not included in drafting drawings. Well, they seem childish for one thing, but in terms of sketching and diagraming, which the chapters in this book will further address, colored pencils can serve as different line identifiers, providing the eye information that can be organized through separation and importance.

Chalk and charcoal are pieces of dark graphite material or composite limestone that offer a sketch quality on paper. The use of chalks and charcoals creates a fluidity of line work, as opposed to drafting pencils and pens for accuracy. These materials would be ideal for the idea sketching process. The sketching process is a creative one and will be identified in the beginning of the next chapter.

4. Pens: ink pens, felt pens, markers, fine-point pens

These images show several types of markers and felt-tip drafting pens used, along with a watercolor set, watercolor pencils, markers, and felt-tip markers and brushes. Courtesy of Ata Asheghi

Pens are quite versatile when used in the drafting fields. A pen is not a pencil and should not be used as a pencil, nor should a pencil be used as a pen. A rule of thumb is that when your notes or sketches can be changed, the pencil is there to rearrange—and that rhymed! The pen is offered on its own, used and depleted when the ink runs out. Pens offer a great sense of fluidity when writing in that the pen can be continuously kept on the paper without any necessary change. Pens used in drafting include felt-tip pens, fine-point pens, and markers varying in tip gauges from thin to thick. Today's products have many varieties, some specifically for detailed line work and drawing. These may be referred to as technical drawing pens. Most of these pens come in gauges from 0.5 mm to 0.8 mm and so on.

These pens offer the versatility of thin line weight to heavy line weight, depending on what detail or line may be drawn on paper.

For the pens that are typically used in drafting and engineering or architectural line drawings, black ink is usually the choice. Brands, such as Staedtler and Dick Blick, offer a unique variety of pens for the designer. Markers also have the means to provide necessary drawing details on paper—not in the drafting work but in the sketching and conceptual phases of drawing. Shop at your local arts and crafts store, and they will be available.

5. Watercolor

Color is hard to mention in the black-and-white world of architecture. This goes to say that what we identify as our selection of colors is offered to the design palette by the ways in which compatibility is explored and used. Conventionally, color and color schemes for the ways buildings are detailed or homes are painted follow matching principles of darkness and lightness, contrasting what is a base to what is a highlight.

The best example might be a hairdresser and makeup artist giving someone a new look. The hair, the eyebrows, the makeup, and the lipstick should, must, or could fit all the appropriate accentuations of beauty. In the construction world, the same should apply. The selected colors represent what a preference holds as if it were something that lends itself to an extension of the self. Watercolor is the best means to create those colors on drawings.

When an artist uses paint, he or she is stroking those lines to make images representational or imaginary with the use of acrylics. Both acrylics and watercolor, however, can have dramatically similar effects of realism when techniques are found. There are many designers that use watercolor in their renderings of their work, with detail on how colors are used and why. Watercolor is a great way to enhance overall renderings when the time is available and the practice is there to allow for reproduction.

6. Drafting tools, rulers, cutters, T squares, glues, adhesives, scissors, cutting boards, toolboxes, automatic machines

Here are some razors and cutting tools, an electric eraser, a flue gun, scissors, triangles, cutting tweezers, engineer's scales, architect's scales, and stencils. Courtesy of Ata Asheghi.

Tools utilized in the practice of architectural drafting, drawing, and modeling are essential to the customization and stock holding of a designer. These tools are found on any retail shelf and can be used in several ways. The tools serve a creative purpose, and when in the use of revision and editing, the right tools can be applied to effectively change and revise portions or parts of drawings. When becoming or being a designer, it is a choice to keep or collect tools as far as what items can be of use to you or someone else in the future. Drafting tools have time value to them; certain times can be used or not depending on the project. "What is the project?" is a question each designer asks when faced with the realities of the times.

Other tools used for design are machine erasers, glue guns, scissors, cutting blades, and X-Acto knives. Rulers, right-angle triangles, and French curves are drafting tools used to advance line work required for the execution of straight conforming lines and curved lines. These tools are appropriate when used in conjunction with the drafting table and for certain tables that do have mountable rulers that swing with the arms. The triangles are essential to create 0- to 90-degree lines, allowing for a dynamic in diagonally drawn lines.

The square triangle serves to create a 45-degree line from the hypotenuse of the angle, depending on which triangles you use, what size they are, and the hypotenuse; the desired result of the degree line will be dependent on that. Some triangles are collapsible, with a center pin, allowing for an increase in the diagonal of the hypotenuse. They are used when stationary, as all triangles operate with a release or screw pin that allows for the swing of the outer hypotenuse to be used to draw diagonals.

T squares are important when a drafting table is not fitted with a square arm or ruler. The T square can be used at the side of the drafting table by holding in place the square holder at the edge of the board or the top of the table vertically. The purpose of the T square is to provide a means to rectilinearly allow for a triangle to slide off the edge for drawing. Most T squares are numbered in inches and centimeters and may serve as a ruler. The T square also allows for fine drafting lines to be made across the entire drafting table in varying sizes, depending on how long the T square is.

Adhesives and glues offer a way to combine parts. For model-making, glues and spray adhesives work with quick application and are used to combine mat board, plastics, or cardboard together. But many types of adhesives, including super glue, may offer advantages in combining parts together for models.

Cutting boards are also in favor with the model maker. They are plastic boards used underneath as a base for placing paper products on to cut with cutting tools. They also vary in size, and they self-heal after cutting, which interpreted means that when a cut is made, the board is not damaged.

Automatic erasers are next. They offer a detailed method of erasing lines, which in drafting plans may need specific amount of erasing done to clear the line without erasing adjacent lines. Automatic erasers come with outlet wires, and some are battery operated.

The architect's and engineer's scales are long three-sided rulers that indicate metric dimensions in the US scale. They are used primarily to dimension drafting drawings in conjunction with the appropriate scale, which would be indicated on a drawing. The architect's scale is used to measure floor plans and wall dimensions, particularly on fine-print blueprints or drafting sheets. The sides are indicated by scale 1/ 3/4 scale, 1/ 3/32 scale and 1/ 3/8 scale, which corresponds to the inch. For the purposes of whichever scale you use, you would want to indicate that you are drawing in a particular scale that converts to feet. For instance, when you are drawing a plan at 1/4-inch scale, you should indicate on the drawing sheet (final sheet presentation, which will be discussed later) that 1/4 inch equals however many feet. For an engineer's scale, the sides indicate ratios of 1:60 or 1:10 to measure 1 foot being equivalent to 100 feet. The engineer's scale is a tool that can dimension large lengths.

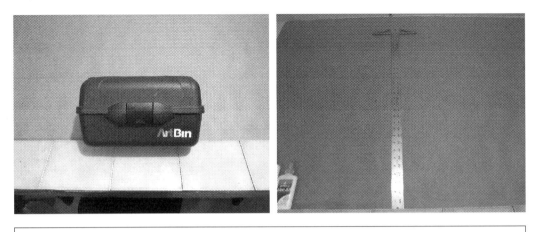

This is a drawing box or art container and a T square. Courtesy of Ata Asheghi.

7. Drafting paper, vellum, sketch boards, and drawing boards

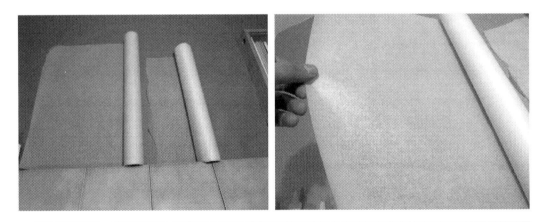

Here are two types of translucent drawing paper, also known as trace or vellum. Courtesy of Ata Asheghi.

Here is a Strathmore drawing sheet binder and a few other drawing and sketching sets. Courtesy of Ata Asheghi.

The design studio always requires the right materials for any development of a plan or project. When interpreting ideas as a designer, the best tool to use is the paper and pencil. It is the tactile method for which ideas are brought to the surface of a visible eye-to-eye relation that affects the senses in many ways. When an idea is brought down onto paper by the application of the hand and pencil or drawing media, the interpretation becomes clear; the user is using tools to express the condition of some idea visually replicated from reality or ideas that are developmental and extend into the vision of the designer.

Paper is essential; it is the two-dimensional surface where, believe it or not, anything is possible. Knowing that the reflection of the self-applied identify of objects and lines can be produced freely gives the designer the capability to incorporate versatility beyond any nonphysical application. The physical application of the act of drawing is a unique way to imagine and create at the same time. This is due in part to the way the eyes work in conjunction with the brain. The paper used is tactile, physical, and blank for the purpose of information. The designer is the utilizer of change, and having the paper in front of the designer gives him or her the freedom to show his or her ideas.

Drafting paper is traditionally a white roll of solid, nontransparent bright-white paper rolled over a cylinder of cardboard. The paper can be rolled out on a large surface and can be cut into pieces. Vellum is another type of drafting paper that is transparent, meaning that light can penetrate through the paper, and other images or graphics will be visible on the other side if placed on top. Vellum is conventionally used on the drafting table and over light boards that reflect light upward onto the surface. When the use of a light board is available, the vellum becomes translucent, and the light can penetrate through, providing ways to overlay and sketch over other drawings.

There are many brands of commercial drawing vellum. Some are white, but most practically manufactured are the yellow trace vellum rolls. These vellum rolls have key design features that will be explored in the strategy chapters.

Other types of paper you may find at art stores are drawing paper, sketch paper, and watercolor paper, as well as a host of others. These papers come in different tones, depending on the weight of the paper and substance. Some are acid free or manufactured without the use of chemicals that would alter the surface friction of the pencil or drawing medium. They will provide good means to produce sketches and ideas or even finished work producible by watercolor or pencil and ink.

8. Drawing roll containers, laminate presentation binders

These are several types of document containers used for presentation and documenting: an 11 x 17 laminate binder, three-ring binders, and a collapsible drawing tube. Portfolio cases are sought after for their durability and ability to hold a series of documents together. Courtesy of Ata Asheghi.

In the case of properly securing or storing drawings for transport, there are specialty items that can accommodate those needs. Large drawings needing to be transported from office to office can use poster containers or cylindrical drawing tubes. Most architectural tubes are for the purpose of transporting construction documents to and from locations rolled inside an expandable tube. This makes perfect transport and prevents damages and folding of drawings. Not only does the container provide the means of setting drawings inside, but some are also fitted with convenient bands for throwing the tube over a shoulder when transporting.

Portfolios require attention to presentation and collapsibility, and by that, I mean that the presentation should be compact and easily readable. For portfolios of presentations, there are several ways designers can showcase their work. One of them is through three-ring binder applications where printed 8.5-x-11-inch sheets can be placed into laminate sheets and fitted into the binder. A three-ring binder is a useful transport tool to allow for snapshots or minimized images of presentations to be stored and viewed for presentation purposes.

Laminate sheet binders are also your best bet in terms of a larger and more presentable means to present portfolio drawings. Many of these come in varying sizes but are typically used for 11-x-17-inch prints. These prints can be viewed by turning over laminate sheets and having clear surfaces for viewing.

Storing and transporting large drawings or presentation boards may be an issue, and for this, art suppliers have been able to sell drawing or canvas suitcases. These containers are like large laptop computer satchels, and they are convenient in application of transporting large drawings that do not need rolling or folding. Some may come in plastic covering in a black color while others are in a matte finish with pieces made from composite flat board. File bags and storage containers also make for good transporting of construction documents, and the choice is, once again, up to the designer.

9. Books on drawing

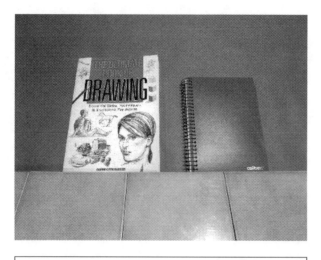

A basic drawing book for learning drawing basics and a notebook. Courtesy of Ata Asheghi.

For beginners, it would be useful to investigate purchasing books on the basics of drawing, the reason being that many of the techniques and applications start with basic drawing abilities. The pencil in hand and the pressure of the lead on the paper are all determined by the skill of the hand and eye with the precision of the idea of the drawing in mind. Many books that give guides to beginning drawing can be found through online shops and bookstores, whereby the knowledge of thin line work and sketching may be essential for the next chapters. Some topics for books that should be recommended would be perspectives, basic drawing, and visual architectural languages of drafting, as well as any drawing book that shows the use of drawings as represented in graphics and technique.

PART II

STRATEGY CHAPTERS: THE PURPOSE OF DESIGN PRESENTATION

This is a discipline. The strategy chapters will explore the means behind how pencils and pens are used together to create what are solid line architectural plans in correct graphic form and organization. In this direct process, the teachings that were previously taught to many architectural students of earlier generations have enabled standardized methods to continue to exist in the fast scale drafting and design process that we see in today's schools. The same schools have taught the practice over decades, with many influential people who have combined both the talent and the conventionality of the Industrial Revolution's late drafting boom.

To mention another path of transition here, for the purposes of older generations and of course the aspiring following generation of prospective incumbents, the transition of pencil to pen when drafting can be noted as a preliminary architectural technique that may allowed in expression of line quality and allowing for correct revisions to occur on plan. This in turn gives more adaptability to producing drawings in pencil and later pen that have immediate interpretations and translations into the computer-aided portions of design.

The next chapters will identify this process. The pen will be used to register solid line work and can be a graphic eye-to-translation method for applications in the design software.

To further mention the historical drafting lineage, which this author has studied, the trace of how the pen was used impacted the final drawings significantly. The blueprint initially became the immediate document of final produced linework and drafting line and., the pen and pencil were the counterpart of drafting techniques of the time which were a way to produce definite results. There is an abundant history of drafting as it dates to modern movements that centered in on the adult man, his table, and his work. The tradition has certainly diminished as a result of the evolutionary development of the computer and computer-aided drafting. To say in addition to what was before, the culture that developed particularly in the United States during the industrial boom produced popular styles and trends along with a pinpoint tradition in traditional drafting customs.

The architect was a select avant-garde elite, a member of a society that had jurisdiction over the idea of image at the time, and to that notion, there were many implications of the image of structures and spaces that transpired over countries and borders, which then overlapped into styles of the times. This is quite amazing. The modern era offered large studio spaces for draftsmen to replicate and create dimension-drafting construction documents as part of an industrialized movement that homed in on the streamlined condition of modernism, expressed by many pioneers, including Le Corbuiser, Walter Gropius, and Mies Van Der Rohe. As part of a movement that developed over a span of twenty years, the international style became the strongest architectural movement that represented the pinnacle of commercial architectural symbolism. With the coming of age, architecture had developed into eloquent themes of esteem and popularity. Architecture was an ambitious career lifestyle with a vast inherency in the fields that allowed for diversity of talent.

The strategy chapters that follow show the perspective of what was before this time and how it relates to what we as professionals wish to do now. These chapters are meant to develop the traditional talent of design and drafting and keep them as a characteristic skill that may otherwise be taken for granted or not used at all with the advent of today's technologies. This goes back to the purpose of this sourcebook, and that is to combine the old with the new and to explore the practicality of senses on the drawing board before entering the media.

One could find comfort and enjoyment in this way, as in the age-old tradition of drafting, once kept for the rulers of boards, the trusted pencil and paper both made smiles widen and pockets jingle. Explore these chapters with your interest and pursue your passion that brought you to this sourcebook, where you feel you may not have found it before or couldn't find it in the first place.

—Ata Asheghi

DESIGN STRATEGY 1: SKETCH IDEAS

6

The idea to implement strategy chapters is based on the premise of allowing readers' learning abilities to extend from basic concepts of ruler and pencil type systems into versatile projections of the hand and eye. This means that these chapters are suitable for people who wish to articulate their rudimentary skills of drawing and line drawing for the purposes of multidisciplinary tasks as far as the design process is concerned. The design process is a culmination of tools and agendas that correspond to tasks, objectives, and ideas.

For the start of this strategy chapter, the fundamental necessity of freehand sketching is documented to show the diverse creativity portions of the beginnings of design. Let us spell this out in easier terms. Say you have a talent for drawing, and this talent is not applicable to anything concerning drafting, design, industrial design, or compositional pieces. You just have a talent that serves to allow you to be creative. This first step gives you the incentive to go further. Now, as the author, I would like to assume that most people who have picked up this sourcebook are of a particular proficiency with drawing in many aspects and can relate to observing life forms and realistic objects to apply the concepts on paper. The strategy chapters serve to address the implicit use of your talent along with the developmental desire to produce ideas corresponding in a manner that can grant you the skills you need to sit down and absorb information. This brings us to the scope of the sourcebook, which is to identify the potential skills you have that can be perfected into a resource available for you to use with your own discretion and your own style.

The discussions on the topics of the chapters following some examples and reiterations will give you better insight into what is essential when stepping into these fields. However, you have learned your skills in your classes or the education has taught you to perform those skills. There are still gaps that were not addressed, which may cause ambiguities between your focus on your subject and your definition as a student or learner. Again, this is said to reinforce a perspective I personally faced many times over that caused an inability to produce and strive forward from the resultant lack of knowledge The word "professional" is so often a word that comes up in life as one that indicates a characteristic or trait of expertise. To that, it is dually important to realize that the word itself stems from aspects and origins of simplicity. For professionals to be considered professionals, much of the focus Is emphasized on the abilities and personages that is exemplified through talent, merit, experience, background, credibility and a well earned host of others. But, there is a fondness of the word that brings individuals closer to obtaining it for themselves and this is due to the notion that professionalism is conceived through practice. Practice that focuses on real-world tasks and behaviors which may be in most cases professional or unprofessional, depending on the quality of the person and his or her actions. Without getting into more esoteric philosophies on this matter, I bring up the term *professionalism* for you to picture it as observation into a picturesque portrayal of how societies utilize practical dimensions of thought versus action in the physical world.

Webster's Dictionary defines *professionalism* as a conduct or aim or a certain characteristic of a profession or type of professional person for the suitability of gainful activity or livelihood. The underlying fact of the matter is what is professional, what one should strive to do to be professional, and what a professional must be is determined by a gainful promise that may consider the term worth the saying. In retrospect, from determining the principle use of the term *professional*, there arises another type of professional that I wish to define. That is the low-professional, which is not a term in the dictionary, nor is it a universal term that means anything. I define the term low professional as an entity that has accumulated what professionalism has reached, however to the gainful extent of being next to what being a professional is throughout a process.

In studio life, the characteristics and personas of people are always brought out by their workmanship, and the constant struggle for inadequacies to be perfected causes friction between the professional and the learner. We have something that is low-profile, which should not be confused with low-professional. A low-professional is one who has maintained the conditions of professionalism all through an age or certain condition, which has not met the bar of truly noted professionalism. With that said, I wish for you to turn your attention to the following chapter on freehand sketching.

A) Freehand Sketching

Freehand sketching is a process wherein the individual utilizes paper and pencil or sketch pencils in a way to create preliminary design concepts and rough ideas. These are usually made with rough sketches that can be interpreted by the designer or by a team. The freehand sketch method is first and foremost a guaranteed method to produce ideas, and it allows for a creative direction to be made whereby initial concepts are realized. The take on this is that the paper serves to allow for the tactile aspect of drawing to take place while giving a sense a freedom.

Say, for instance, you see a building or a place and wish to take ideas from it and put it on paper. This takes quite an imagination, but it really comes down to the focus on how you feel about drawing. Drawing is an enjoyable process; it is an activity you can find solace in by putting on paper whatever comes to mind. It is a resource, however, that may allow you to source and repeat through the hand the essential qualities that make up a drawing. Freehand sketching is a unique method that gives you the means to draw without discrimination and without constraints. Further, the freehand drawing process is tamable, meaning that you can grasp the idea of placing a pencil on paper with ease and without frustration. This does not limit someone to the possibilities of being able to produce images without a determinant reason for doing so. The craft in the freehand sketching process is where one can assess his- or herself for the versatility of the line. The freehand sketch is the opportunity to enlighten your capacity to produce without anyone telling you not to. Or in the case of working or sitting with a team and producing drawings, the feedback would be a good way to articulate the sketch, whether it be a small drawing or idea or a piece of building or what have you. The sketch is performed in a way that shows the uniqueness of the person's mind on paper. Here to say again, that when you start a freehand sketch, you are focusing yourself on what it is you wish to see or what it is you wish others to view. Keep in mind that it is a developmental process.

Freehand sketches are loose and rough and usually allow for the line to be drawn in a manner that proposes simple concepts, ideas relating to form and shapes and, above all, the parti. The parti diagram is a form of freehand sketching that incorporates the use of simple diagrams and identifiable illustrations that respond to concepts unique in the process of delivering built ideas. The freehand sketch is a solution for the first implications of how to process activities on paper with definite assumptions as to where ideas are developed and how they are developed. The sense to the direction of where drawings are first made comes from the ability to perceive an idea and then to incorporate the characteristics of that idea through meager rough drawings. It is easy; it does not require critical thinking or obtuse reasoning, and it follows a discovery method that is based on multiple assessments of line and form.

When you perform a freehand sketch, you are indicating the line or the loose line as a way to create form and place. You hold the freedom to assess your drawing based on what you wish to see and what you wish to accomplish. The greatest abilities of drawing are unique to the understanding of obtaining realistic snippets of the surrounding life you are accustomed to and that you bring into focus on paper. So, the only gap that many people fail to address is the notion of emptiness on paper. To describe the sense of emptiness or having nothing drawn on paper is to assume that there is no ability to create. So, when you possess the ability to create a freehand sketch, you are putting your assessment to work in front of you.

The freehand sketch allows for the versatility of decisions to unfold, and it allows for the capacity of dimensions to reiterate into actualities. By taking a freehand sketch, you can form a generic or idealistic idea into a picture and develop those ideas further with the aid of notes next to the sketch or notes associated with the sketch.

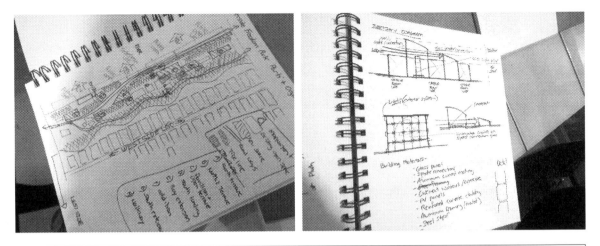

Freehand sketching is a developmental process, and it goes hand in hand with simple paper drawings. When starting a freehand sketch, try developing one idea at a time, and then through several other pages, develop the sketch further. You could use notes and little numbers. The sketch should be in *pen* and can indicate the abstract or definite level of image you are trying to produce. The level of clarity is dependent on how you use the line, and in this case, from my selections of sketches, there are diagrammatic sketches showing the use of line and intricacies of detail. It is interesting to detail, and detail provides information and quality. Courtesy of Ata Asheghi.

B) Trace Overlay Sketches

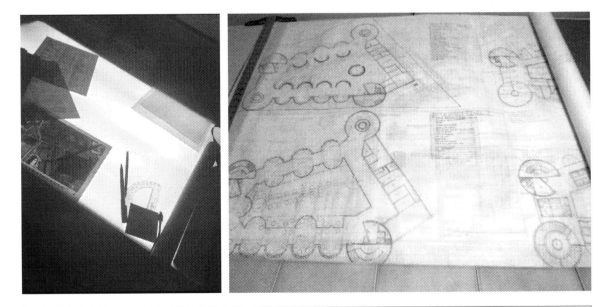

This is a light table, which is used to overlay drawings and templates with the use of florescent light. Next to that is a large-scale trace paper design of a building on a table. Courtesy of Ata Asheghi.

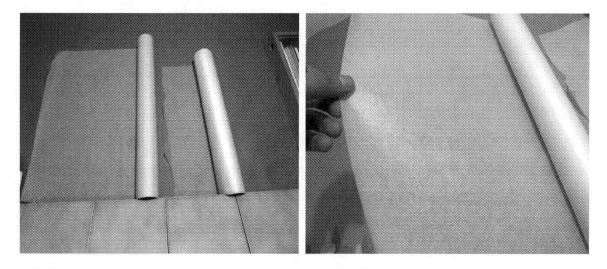

Again, the trace is versatile and should be considered in conjunction with a light table. The trace provides the ways to overlay and examine the drawing by overlaying lines onto each other. Courtesy of Ata Asheghi.

The tracing process is a special and useful way to allow for visibility to occur on paper. This is a process where translation is effectively brought to the senses using vellum or trace paper. Drawings that are overlaid one on the other and that are used with light as a background source can be traced over to other drawings and in this way provide visibility when translating line over line. This is suitable for when the actions of producing line work require an underlayment of line visible to draw over and perceive. The task requires that the pencil or pen be used over the trace and that there be an image or underline image behind the trace. These drawings are made by the attempt to overlay and assess the drawing from a layout perspective. This gives the reader the ability to postpone any developmental improvements by making time to check and review the underlayment image. The trace serves the designer as a tool or method to correct and reevaluate the line in terms of its correctness. Don't worry; it gets better. It's not too deep into the pond. Not yet.

Tracing is used in the office environment as an architectural keystone method, which may also be acknowledged as a fundamental drawing method, giving the ability to show limitations in drawing and how they correlate to actual designs on blank paper. I wish to say as the author of this interpretation that the line is important to address in terms of its clarity. Architects tend to fumble around drawings and at times go off on conjectures that are not easy to replace with standardized corrections. The notion of the structure being bound by constraints and conditions leans closely to the objective of a form. But when there is disarray to those forms, there can't be any equal condition to what should be as opposed to what could be. So, in saying this, the process of tracing provides the means to reflect on the actuality of a design image as opposed to what can be directly visible without readable proof, that being the image behind the image.

Here are a few steps involved in the tracing process that will enable you as the designer to incorporate your sketch ideas onto trace and allow you to replicate drawings over each other or onto your large drafting sheet. Keep in mind the tracing process involves erasing and checking for line clarity and accuracy.

1. Gather materials needed for a trace outline, including a pencil, white paper, blank paper, a sheet with an image or site plan, trace paper or vellum, a ruler, and triangle access to a light table.
2. Start incorporating your sketch onto the trace by having the sketch idea under the trace paper or image that you wish to trace over. Place a piece of trace paper over the image or diagram and tape the edges.
3. Now, with a pencil, start to draw over the original drawing or diagram and include a border around the trace.
4. You may need to consider removing the trace from the diagram often to see the image behind and include the line over the image correctly using your tools.
5. Once you have produced a trace drawing, save it and keep it filed for any use over large drafting drawings, which may come in handy.
6. The last thing to note is the ability to use pens, markers, or other suitable drawing media that would help identify the trace through notes or line ideas.

Tracing can be used extensively throughout the design process. It is a skill that can be used in the office or on practice light table. The emphasis is to produce drawings that are traced over other drawings and in this process create separate trace outlines that would be compatible with larger design works.

C) Concept Drawings and Interpretation

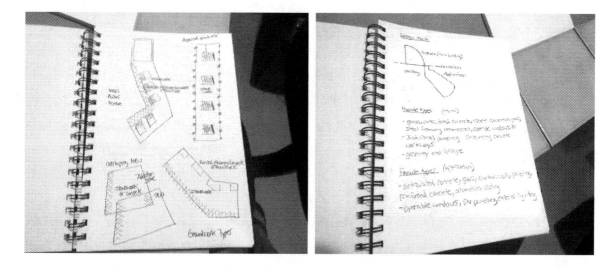

Concept drawings are a variation of a sketch produced to provide information, form, and special conditions. The concept drawing is a longer process because it compiles information and sets of drawings to produce a defined single concept or multitude of concepts. Your concept drawings should be objective and should show a level of dimension and objectivity. Not to mention that this would be a piece to your presentation in a more visual and comprehensive sense, which will be discussed later. Courtesy of Ata Asheghi.

The conceptual phase is determined by the originality of an idea and the originality of the development of the idea through concept means and through gainful insight of how a concept should be created. The sketching process utilizes the hand-eye skills of drawing, and when done with consideration of line and graphic quality, the result of the sketch becomes a concept. When a concept is born, the characteristics of the concept and its quality begin to envelope the sketch, and then using notetaking and color or shade, the concept becomes clear. What is a concept? Without getting further into the detailed aspects of what a concept is or where it is derivative, the basis of the concept is the generation of a perceived visual and environmental idea that is viewed through the designer and projected onto the paper. Yes, there is not a true explanation of where a concept comes from but looking closely at the self-examined persona of architecture, we can determine that an assumption can be made that concepts are creatively shared, intangible ideas. To further this into meaning for readers who are translating to the fact that many of these designers lack the impulse for a creative outlet. To challenge this, one must find him- or herself in a certain demeanor that would enable a constant flow of ideas to generate on paper, thus giving the sequential number of ideas replicated or produced to be a single concept.

Concept design is a formula as will be described here. The term *concept* relates to a stem or a source of a realistic item or object perceivable in nature and acknowledgeable by man. The concept design is the momentary formulation of what that item should, could, would, can, may, might, and will be once conceived. The concept design involves a set of parameters that are not in the form of a list, although it could be generated in those means if involved with an academic assignment. When designers think about concepts, they run into obstacles that limit their scope on what is a defined object or item. In terms of how a concept is formed, let's take an industrial designer as an example. This will give a better insight into the formula process involved in producing a visible and documented concept. The industrial designer wants to create a new chair that leans back on itself with the control of a button or switch. The concept is clear: there must be the design of a chair and the design of an electronic switch that would lean the chair back. The design of the chair is left up to the designer, and the overall structure and mechanics are developed through phases of design. The concept again is to have a chair lean back with a switch that operates on the hand rest. From this concept, the following determinations must be made to accomplish the task of evolving the concept. These are in step and in principle as an outline to any conceptual design as a formula for creating the concept:

1. The perception of the concept (What is the building or space?)
2. The stated idea of the concept in clarity and in defined characteristics
3. The concept sketch that is developed through freehand sketching
4. The development of the sketch with tracing or multiple sheets.
5. Collaboration or individual reflection on the concept
6. Acknowledgment of the concept as being defined as the stated idea

For all intents and purposes, there is a figurative method and a formulaic method to configure a source for a concept. By the steps previously mentioned, we examine a formulaic method that is not often determined by any certain constraints. Thus, the need for a program is derived to incorporate those specifics that are essential to an "outline" per se of creating dialogue toward the development of a concept. When the formulaic presents itself as a possible guideline for completing tasks, the result may be included in an academic or professional environment in which the development of a concept is explored thoroughly and with prerogative. The figurative method that will be mentioned here is a way to respond to drawings and ideas loosely and by the comfort of the pencil and paper, something that is explored often by detail and response to the visual senses.

D) Patri Diagrams and Concept Diagrams

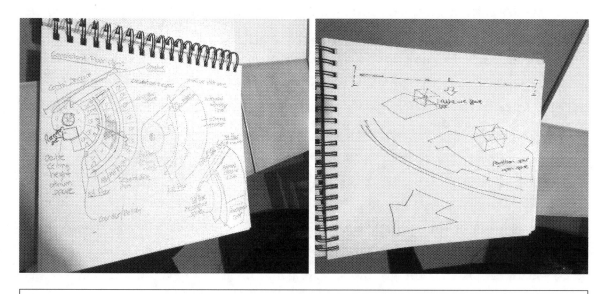

Parti diagrams are graphic interpretations of what you wish your *building* or *space* to be. Courtesy of Ata Asheghi.

The parti is a way to combine special features, volume, and line all together. When drawing a parti, first think in simple volumetric terms, such as a square space and circle space together, or what have you. Then work on them in a series of consecutive diagrams called concept diagrams that show where and how the spaces will lead into being a progressive building or space. The concept format allows you to control the direction of parti by indicating:

- **Egress** (the pathway or direction of entrance or exit)
- **Procession** (the sequence of movement through an organized space)
- **Volume** (special condition that represents a form or composite form of a proposed actual space)
- **Shape** (the basic shape of the form—circle, triangle, square, rectangle, oblong, octagon, hexagon, pentagon, decagon, oval, curve, stair—type).
- **Form** (a way to create the spaces or volumes in a way that expressively shows a creative form or structure)
- **Function** (The intended use of what is drawn and how it is to be used in the actual terms)

In response to the direct linkages between conceptual and observational drawings, there are two methods to differentiate between what can be represented as actual versus what is perceptual by nature in terms of the drawing. In our studies, we take two forms of preliminary conceptual design, and we put them into what is called a "parti" or a "parti diagram," which, in expressive terms, represents a conceptual drawing or multitude of drawings that offer a dynamic set of images in relation to form, scale, objectivity, subjectivity, and expression. The parti is developed to organize the principal ideas revolving around the development of the concept and the utilization of line to fit ideas into a useful tool. Taking this further, the adequacies of a parti diagram are based on the content of information being provided. So, the difference in the parti and a concept diagram is vast.

Lastly, the parti diagram is an extension of the freehand sketch, which can be formed by certain geometries and shapes, such as consecutive circles or squares that overlap each other and offer space that you can write in. It may be also used as a presentation diagram that can show volumetric shapes and scales of what is to be visually drawn (as far as buildings go).

The concept diagram is separate from a parti. There is no need to mix the two together because they are separate but equal, and they hold the same technicality as a freehand sketch. The diagram involves a rigorous set of ideas to be produced in way that incorporates different dimensions of a unified and collective processing of information that offers suitability to the designer in the use of itemized drafting via drawing tools. The concept diagram consists of notes, usually in pen, and contains corresponding information to what is drawn and what can be explained. It is simplified into lines and notes but is not restricted in presentation and format techniques. The concept diagram should not only aid in the discovery of utilized themes and directions of design but should also be a befitting technique to materialize environmental ideas into 2-D diagrams. When there is also a consideration of modeling, not for *Vogue* magazine, but for the 2-D drawing, we must assert the environmental variables into a cohesive image of what is to be shown.

DESIGN STRATEGY 2: CONCEPTUAL LAYOUTS

7

We have examined the sketch idea process whereby the initial phases of the design sequence start. Through sketching and sorting conceptual diagrams into the picture, a better sense of direction comes into perspective. The next phase to address would be conceptual layouts and how different drawing techniques are used to articulate the ability to draw and design in this strategy chapter. In the previous subset topics under the first chapter, I described an introduction to freehand sketching, and now, as we investigate the finer details of the technical productivity of drawing, we must explore a more precise look at the drawing skills. With this, the following chapter after this will get you right into the fundamentals of hand drafting. Take into consideration the work you have already done and the amount of time it took you to produce your conceptual sketches and napkin-style drawings. These are the kiss after the tears, and they will grant you the necessary means to develop them further.

A) Freehand Drawing

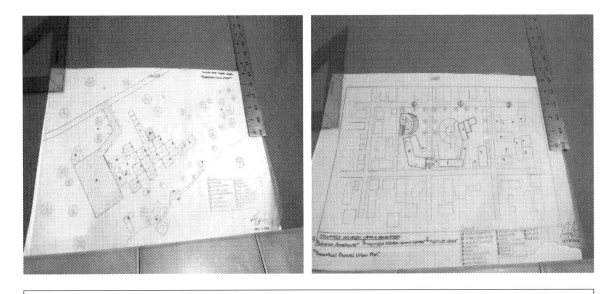

The freehand drawing process is the way the pencil and the ruler are used to produce more accurate drawings that can be interpreted for the purposes of producing later drafting drawings. These drawings are suitable for document control and document analysis. The drawings are meant for your consideration in developing the next drafting phases. Use these for when you are sitting by your drafting table and looking over them to transfer ideas onto the sheet. They would not be direct drawings of what you wish to put on the sheet but transfer drawings for your process of drafting. These are a few site plan drawings. Courtesy of Ata Asheghi.

Freehand drawing is an irreplaceable method, as opposed to sketching, in that the process involves a more finite response to the level of detail required to produce more accurate images. The freehand drawing must present itself as a complete and comprehensive image that works to fit the need of a reality-based image. This is said because images produced in artistic terms must provide an actual sense of something as opposed to drawing doodles and happy faces. Now, I like to assume that those of you who read this can draw those doodles and happy faces but wish to amend themselves for their legitimate talent of drawing, which comes from our childhood.

Drawing is an enjoyable process, and it gives us so much to know how to produce drawings. Whatever is to be drawn makes sense only to the person who is drawing. The talent of drawing is one that can be refined—and I say can be refined because it reflects the age and talent presented by that person at a particular time. What this chapter wishes to show is that the freehand drawing process requires attention in order to capably associate your drawings with realistic environmental or architectural themes. As the author, I provide this information to those who want to understand the developmental aspect of how to produce drawings and what a good drawing can encompass.

Art class is a scheme. It is something that many of you are used to from what your teachers have taught you to conform to. That is not saying that you do not have the freedom to produce what you like but rather what you like to see better. In this, the opportunity to explore your talent will be tested by how well you can take your time to again enjoy this process. The freehand drawing allows one to excel into more vivid and achievable goals, goals that can offer a response to your talent and your finances. Other than that, your drawing ability and whatever you prefer to create could be subject to diminutive or even embarrassing criticism. To say further, if you consider your talent to be miniscule to what other talents are or were in the world, then there would be no rivalry to your advancement in producing equivocal or similar works. That is then to say that your drawings can be workable and can be of worth. The humor, of course, is like a bad comedian who just wants to please the eye from something already so visible and objective that it only sparks the sense of its own stupidity. Take away from that and make it so that your talent and your enjoyment strive in the sense of your critique, which will be developed by yourself first.

Many people might read this and wrestle with their frustration at why a guru-type methodology is being analyzed here. This is not a soap opera, nor is it a one-on-one counseling procedure on how you can learn to sharpen a pencil. This is for you to demonstrate what many cannot perform. And this is for you to generate the application where your ideas lead you. With that said, let's dive right in.

A) Freehand Drawing

Let's now bring into our perspective freehand drawing and see how that can have implications for what is perceivable and what would be guaranteed in the sense as a technical document for drafting purposes. The aim here is to register the idea and introduce freehand drawing as a solution or system that may be considered for the phases of drafting and document sheet production. As the design quality develops further from the sketch to the concept design, we need to note that the single most important asset to you would be your ability to alter your freehand drawings and tailor them to fit the later corresponding conditions of a design phase drafting design sheet. I must say that you have reached this far without any preconvention to what you may have thought before you were introduced to this. But now you may be recognizing that this is a sequential follow-up to the preceding pattern that you yourself wish to develop and commandeer in demand for your own goals. To reiterate, these outlines are provided for you to maneuver beyond your own expectations and give delivery to the criteria needed to further the design advancement process.

Freehand drawing does not replace sketching, nor does it replace small drafting processes. I mention small drafting processes because there are ways that drafting could be done as a drawing ethic on a drafting board, which limits the larger capacity to draft sheet in full scale. This will be examined further in the drafting design chapter, hopefully to cross the road of no direction into more visible terms. To identify freehand drawings being the primary delivery method in this process is to say that the prerequisite of your concept has reached its dimensional level.

On a 2-D surface, you can now envision your concept fluidly and give more insight to the readability of what you are trying to conceive. It is not because of your inexperience that your skill cannot be adequate to continue further in this process. The skills acquired through application learning bring the only suitable means of knowing what to do into fruition. There and there, you must learn to take what is in front of you, and you must quickly adapt to the useless demeanor that may be presented to you, placing your skill in refinement to allow you to progress over the constraints of time, place, and person.

The freehand drawing method starts with your clarity of decision, in which you set forth what you wish could be a visual illustration of your own knowns. Then consider the unknowns from a perspective that relates to an artistic function for your own expression. The intensity of not knowing what comes next is not an option for you, as you should already be accustomed to holding the sense of things close to your intents. So, in the freehand method, your drawings may serve you as a replica of what it is that would be final to you. Here are some ways to accomplish this idea. First off, bring your sketches and your concepts together and look over them. It would be convenient for you also to have a notepad around to jot down what your improvements would be or what your task requires you to fulfill. Over the common sense of things, including certain notetaking applications may offer you an edge toward refining your concept into a workable drawing. This is a work in progress, and, as the author, I would emphasize greatly this dedication to the working process, which may be a concern of employers seeking to compartmentalize applicative tasks. In engineering terms, this would be a credit to your own craft. Develop the drawing to the best you can and admire your pursuit, which may offer you inclusiveness toward your final. Your drawing can be a mirror image of your own self-reflection, and I say this to fulfill the aspects of a versatile talent.

B) Drawing Criteria

Producing the drawing(s) should enable you to further develop your concept diagram. The process should give you the opportunity to demonstrate your detailed level of work. What you should be looking for in each drawing when creating a scheme for a building would be the level of detail. Here are a few of the following types of drawings that may be performed in advance before the drafting phase:

Initial site plan drawings

After gathering information from the site via maps or online map apps, take a print of the desired location at a desired scale and mark over the site of the image in pen. The markings should indicate your desired building plan level with a surrounding line that fits the parti. The markings should also be made to indicate direction, sun path, orientation, foliage (such as trees and plants), hardscape or landscape considerations, and entrances. Then overlay the image with trace and produce the drawing or several series of drawings that will offer you a more developed look at the site. Your best drawing should be a finalized version of your conceptual site plan indications.

Building concept drawing

After developing the site plan, gather your sketches and concept diagrams together. Gather as much information as you can from these, and take notes on what you wish to include for your total building concept. As a hint, your building concept should be relatively easy to produce now that you have an idea of how to produce the site along with the given form and detail of the parti diagram. The information you provide in the building concept drawing should all be comparative and should be a work in progress, including tracing and replicating more ideas onto a final level. Things that you should include in this drawing would be the dimensions given to you from the site and from the scale of the site that you are incorporating into your building concept drawing. The dimensions should reflect accurate numbers in feet while being conscientious about the appropriate conditions of the building around the site. At each line you produce for your building, annotate or note the dimension to scale of the size or length of each part of your building. The building concept drawing is a general document that shows building level, form, and size related to the surroundings. Sun path and north light, as they are called, show a direction of the sun from which ever longitudinal direction you are. They are produced by orienting the building to the side where the suns directly impacts the building. Then start detailing lightly your mass of the building with notes and considerations for future drawings. This is a direct conceptual method that creates the overall appearance of the building in terms of mass, proportion, dimension, angle, relation to surroundings, and feasibility to site.

Floor plans and special volumes

After you have produced the initial building concept drawing, take from that all that you have noted and start to lay out your building floor plans. The key here is to observe the working process that you have started from the concept to the concept formation and allow for creativity at the plan level. Floor plans should indicate dimensions of wall lengths as well as basic sizes and square footages of rooms and hallways. Here is a list of spaces that buildings have in the architectural fields: rooms, hallways, foyer, lobby, W/C or water closet space (restrooms with stalls and urinals, showers, sinks, and vanities), meeting rooms, conference rooms, activity rooms, mechanical and plumbing rooms (e.g., furnace room), office space room, subfloor or basement, storage rooms, laboratories, classrooms, lunch rooms, computer rooms, custodial closets, closets, balconies, upper-level double height rooms, stairs (indicate length and rise over run, which is typically a seven-inch rise and eleven-inch run), shower rooms, locker rooms, gym rooms, basketball court, tennis room, supply rooms, bedrooms, living rooms, family room, kitchen space, garage space, shed or supply room, gathering space, front desk room, presentation room, laundry room, decks, patios, porches, studio space, gallery room, retail space, mixed-use space, retail store front space, bars, lounge room, club gathering room, crawl space, docking area, supply closet, fitness room, yoga room, family activity room, playroom, children's room, filing and paper room, corridors, and attic space.

Along with the types of rooms, you need to indicate certain features for the plan. Here are some building or residential features you may need to add: windows, doors, operable windows, bay windows, consecutive windows, vanities, showers, desks, chairs, countertops, furniture, lighting, shelving, electronics, carpets, rugs, garage doors, basement doors, sliding doors, cabinets, kitchen accessories, railings, posts, basement windows, filing cabinets, sculptures, plants, animals … Put 'em in there!

Floor plan layout techniques

Incorporating certain techniques for the working process of the floor plan layout really comes down to your ability to center on your conceptual idea. From what is derived from the concept, it should give you the incentive to start your floor plan discovery. Exploring a floor plan layout of a building would both help you explore the potentials of spaces and would also engage you in identifying how spaces are formed together architecturally with certainty to use. Presume that the spaces you wish to control in the layout are multiple, so you must take advantage of the freehand drawings to govern your decisional process needed to fit the requirements of the building you have formulated.

Other factors that may be involved are external, meaning that quite a bit of research would be involved for you to gather ideas about your development. For one, as a reader, many of the books you might have read on architecture should show or indicate plans that may give you a good look into what your proposed scheme may be. Taking a research method approach to this before your initial floor plan layout will ultimately give you an architect's mindset and will allow you to discover solutions to spaces you wish to introduce in the plan.

Documentation and Preparation

This is a developmental process that would involve a separate conceptual diagrammatic drawing inclusion. This involves the proper documentation and application of certain working drawings. Try to sketch out several floor plan drawings together, take notes, and annotate dimensions and form. The premise of this is for you to explore the potentiality of spaces and new spaces and how they would relate to each other in the building. It is a tricky process, and it is one that you must understand through human anatomical and physical function. I would recommend looking at architectural residential and commercial design books that show plans. Try to examine them and understand how the spaces relate to each other. Through your defined examination and observations, you will gain a better attempt to configure spaces that are part of your desired preconceptual phase. The relation of spaces and how a designer configures those spaces in separation and in consideration for the overall function of the building is determined by these principal design objectives

Design objectives for determining spatial relationships and feasibility

1) Determine the use of the building. For the type of building that you are engaged in producing, there is a continual mention of the concept as it branches down into all aspects of the spaces. The concept and use of the building must match the special volumes you are integrating and must also bring to attention the versatility of those spaces throughout the design process.

2) Negotiate primary spaces, secondary spaces, tertiary spaces, and so on. You have to explore the dimensions of your ideas and concepts in order to value the process needed to associate spaces to each other. Take for instance how a flower looks, a center surrounded by petals and a stem. There are three components to a flower, maybe a fourth being the roots. Segment the flower and understand the relations between each part and how they all fit together.

3) Determine form, linear form, abstract form, geometries, divisions, expression, and conceptual form. When you are producing a building or structure, there must be equivocal assumptions made about how it should look. The visual appearance of what you are producing should again relate to the concept. The concept diagrams have outlined for you your response to the use of the building and what you intend on having the building cater to. The design intentions are configured by how you decide to plan

the spaces according to the site constraints along with your preliminary view of the shape and form of the building. In plan, you must examine the potentialities of spaces being part of other spaces and how they are executed in the design process. You must consider why each space deserves its location and how it relates to the direct form or entirety of the building mass.

- Linear form shows a direct single or axial pathway that indicates singularity, line, consistency, and a direct path.
- Rectilinear form indicates a square type or boxed type cross-axial pathway, geometrically bound by the shape of several squares.
- Circular form shows a dynamic of circular objects and corresponding forms that are representative of spheres or a range of circles.
- Diagonal form shows a representational asymmetry of form that may be created to show a diagonal or angular direction.
- Triangular form shows a representation of a three-sided figure.
- Single form shows one single volume or form.
- Double, triple, and quadruple forms are extensions of one form.

4) Create a site plan recognition and building orientation

The site under your overlay that you have printed or acquired should give you the best idea about your building or residence. The site shows the geography and the total available surface area for you to build on. First, we need to say that the site is the space to work in and the space that you will organize your idea in. Zoning requirements are certain conditions set in place by the town or city where the site is located that indicate to the builder the specifics of the site in relation to its surroundings and what the site contains and does not contain. The zoning requirements show the linear square footage of the site; the footprint of the site, which shows each perimeter dimension, and the category of the site being a designation for a buildable area or other use. Some indications from cities that the site is under a 2B site zoning regulation or other criteria that fits the municipal categories for a piece of property. Zoning designations such as 2B or Z or R would indicate the occupancy and appropriate site conditions for allowing building.

When you are designing the site that you are working on, first recognize the site in its entirety as you did with the diagrams and conceptual phases. The drawing overlay of the site should give you a determination on the actual design form and layout of what the building looks like from a bird's-eye view. By incorporating your drawings, sketches, and concepts into the site, you are recognizing it as being actual and a part of the surroundings. Make sure to indicate on your site diagram your zoning dimensions if they are available to you and your north direction by drawing or placing a circle with an arrow pointing north at the corner of the drawing. This will be used in drafting later. For the purposes of the next chapter, this is a detailed explanation of your preliminary freehand drawings of the site that you will use to develop into a program for your building. In part B of this chapter, we have explored the drawing criteria for your spaces and preliminary floor plan ideas for your intended building or residence. Again, the next chapter will show you how to form the program for which you are developing this into a whole with integrations into architectural styles, designs, themes, and building types.

Orient your building on the site drawing so that you indicate direction from north, south, east, and west. Make sure to draw clearly and visibly with the use of your ruler and your curving abilities with your wrist and pencil. Two-lane roads should be lightly and loosely drawn, and major roads should be drawn with a ruler for direct line work. In the case of the urban framework, a great tool to use is the grid. Organize your drawing of your site and building on paper and measure a crosshatched grid of lines on the paper corresponding to the grid that you are working on in the city or use the grid as a measuring tool for the longitudinal and latitudinal measurements for your building. This will help you understand how to work around the grid line and give you ideas about your forms and the volumes of your building.

Another good method to use on the site orientation is the site plan freehand drawing with notes and a key next to the edge of the paper. The key can be made to identify items on the site drawing like trees, which are drawn as a circle with lines coming out the middle. The key can be introduced to help navigate the site and allow the reader to understand what is on the site. This also helps register the geography and the objects that are visible to the draftsman for designation purposes. A key or site key can include a box with small graphic symbols and next to those a description of what they are. Here are some site key or legends as they are called: property line, building line enclosure, fire hydrant, tree, bush, fence, axial, cross-axial, sunlight, north light, north, south, east, and west.

Also, some other uses of the legend can be employed in component-type freehand drawings of the building showing the same dimensions and layout, using trace overlay methods of the building. These may show materiality, what the structure is, or what each part is from a bird's-eye view. The legend gives the view of the direction correlation to items on the visible sheet.

5) Elevations, structure, sections, and 3-D drawings

The other method to visibly interpret freehand drawings is in the inclusion of elevations, sections, and 3-D drawings. As you already know, the floor plan and the site plan show the graphic overview of the building spaces, as well as the overall detail of the building or residential composition. These give the reader or the designer the ability to understand the volumetric dimensions of the spaces and how they relate to each other as part of a whole composition of design spaces. To show in further detail the image of the building in relation to proportions and sizes, one must examine the way an elevation is drawn.

An elevation is a side view of a building where all aspects of the plan are interpreted on a raised surface of the building. The elevation shows the detail and building facade type using height. In a presentation, four types of elevations are shown, and they are the north, south, east, and west elevations. When these are shown in a plan or when they are shown in relation to the floor plans, they are considered as exterior examples of the project you are developing.

Start off by laying down some trace or vellum on a surface and taping the edges. Then bring in your floor plan drawing, which should have dimensions corresponding to the length of the sides and features. Then determine the scale of the elevation drawing in respect to the floor plan dimensions. Indicate the scale in relation to the plans, and draw accordingly; for example, if your floor plans are at a 1/34 scale, then assume or repeat the same scale on the elevation. Then, draw a solid darkened black line near the bottom of the trace to indicate the floor level. If your floor level has a hill or a slope condition, draw the solid straight line, and then draw a crosshatched line indicating the hill over the straight line. Mark the straight line with points or lines showing one end of the side of the building to the other end and the walls separating different sides of the building. You should end up with a piece of trace paper with a straight-line showing marking on the line for portions of the side of the building or residence you are working on. These markings will allow you to work on the drawing and include a sense of height to the proportion of the side of the building you are working on. Designate the height by drawing a line vertically and deciding the proper maximum allowable height of your building. A code-compliant standard for ceiling heights in residences is usually seven feet six inches above ground. If you are designing a larger space, there must be considerations into the materiality and structure of the building.

For large spaces that require higher sides, determine what components make up the building to allow for the appropriate height. For example, warehouses consist of steel girders and large frameworks. You might need to reflect the height of the structure or framework components first by indicating your total height. This would be a separate drawing or schematic, which will be explained here.

Structural Freehand Drawing

With everything stripped out of the building first, you then must look at the simple components that make up the framework of what stands. Structural steel and wood are usually used as the skeleton of the building and are planned before any facade system is placed over the framework. If you are currently studying or taking classes in structure or statics, you would have to recognize the precision involved in creating a structure from scratch. These courses are detailed, and they home in on mathematical and arithmetical procedures to calculate building loads on the building supported by the structural system. For the sake of the design practice, these will not be introduced here; rather, I will offer an objective examination of how to incorporate structural qualities before determining the facade and envelope or building skin.

Identify the structural components you wish to integrate into the building by drawing a freehand callout sheet of materials. These should include a list of the components you wish to include in your overall design. The list should show the detail line of materials that are conventionally used in conjunction with the floor plans and elevations. Determine what parts are needed in the building and what parts would qualify to be used to set the overall building heights. For example, in the case of structural steel, try to include your framework, posts, and columns on plan. There should not be an analysis into the mathematical determinations for structural loading of live and dead loads rather considerations into the stable framework of what is being developed. Take what you have from your callouts and draw a basic representation of the framework of the building. By doing this, you are allowing yourself to understand what is needed in the proportions of the building. Designating the columns or posts gives you the certainty of knowing the visible conditions that make up the structural aspects of the building.

A section is cut through a building, showing internal details. The section is very helpful in understanding what is called a diagram Poche, which shows the resolution of a built solid and unbuilt void system with graphic visual outline. The Poche is an outlining method that articulates the volumetric conditions of the internal special features of a building. When introducing the Poche, there must be a determination on the section of darkened lines that correspond to the outline of the special walls and features of the building. In looking at this technically, the objective to produce a line is in the dimension of how the section is removed from a part of a building and how it is recreated into a feasible and entire composition in what is also called a sectional diagram. In the architectural world, these diagrams are used to conform spaces together and help administer the function of special use.

Again, without getting too explanatory about how this relates to the presentation aspect, the drawing itself serves as a visual reference to allow for a dynamic expression to be made on a two-dimensional surface for the purpose of exploring the volumes of the space. When drawing a section, pull out a dimensioned box from the building. You can do this by cutting a line down the center of the building or at an angle and drawing what is interpreted from a chosen side of the building. This gives the reader a view into the internal perspective of the building while it is being articulated into a graphic that can show relative heights, people-occupied space, and the building's dynamic structural features. If you can produce an elevation, the same principles apply to the section, and you may admire what you have produced.

Some key highlights to incorporate into the section would be the use of a color scheme to address intricacies or details being made with notes. The Poche of the wall lines could be drawn in two ways with either solid black lines or outlined walls, whichever you prefer. The objective is to produce a two-dimensional image that reflects what cannot be seen inside. Articulate the preliminary drawing to add into the final presentation drafting sheets by including scale, dimension, sunlight path, and materiality concern. The materials you used from the callout should be included in the section and should be annotated with identifiers as to what they are and how they are incorporated. Take your time to enjoy first what you are doing and make clear definition into the parts of your spaces. A section should look like you have cut open an orange and laid the insides out flat.

A 3D image shows composition, arrangemnet, form, shade, volume, structure and line.

3D objects are multi-dimensional representations of 2D lines put together to form a volume object.

The composite in this image shows a volume of a square with shade and form.

Using 3D forms can show the direct dimensions and conditions of line and the use of spaces involved with each other.

A Three-Dimensional Object in Space. Volume and Form.

A 3-D drawing may aid in the process of building recognition and building dimensions. Also called axonometric drawings, these drawings serve a multitude of purposes. The 3-D drawing is a great way to show the viewer a composite view of your building or residence. It would take some skill to create this, and it requires several steps involving perspective and relation to plan. Come up with an idea to organize what you wish to see in your 3-D drawing, and sketch out some variables on a piece of trace. They should show a volumetric depiction of what you are trying to develop as a 3-D image, showing all sides and lines upright and in an objective form. Start off on your sketch drawing by drawing a perspective line from a single vanishing point or two vanishing points. These are two points on the paper equidistant from each other (having the same linear distance). Draw a light line connecting the two. Then mark the center of the linear line. Take the center line, and draw a longer line on that mark, going perpendicular to the linear flat line. This would serve as your reference line for your three-dimensional building. Connect the top of the line and the bottom of the line with light angled lines attached to each of the two points. You should come out with a perspective drawing that shows a single visible center line and separation. Make sure that the height of the center line is to scale from what you wish to produce. Then work on the line of your building or residence from the connected separation lines.

In order to understand this further, you may wish to bring your elevations closer to you so you can see where the sides of your building are corresponding. Draw out the lines from the center line, showing the different parts of your project. You may also wish to add heights to the center line. Once you have produced a basic geometry of the shapes of the volumetric drawing, you can now transfer them into a freehand drawing with more detail and accuracy.

6) Create a building callout or multiple view set of details of the building

You may run into a design gap, as I have expressed before, in how to relate materials and ideas of your building on paper. These may be stuck in your head and may frustrate you. You have to figure out how to provide this information to another viewer to make sense of what you are intending to use or show out of your building. The callout or zoom image drawing gives you the best idea on how to graphically show what each part of your building is. Start by jotting down a list of materials or items you wish to place into your building. These materials may have been selected from a magazine or a building specifications materials inventory list. Having these items noted is crucial for what would be called a callout list. The callout out list is a drawing that shows separate components and materials selected for your project. They may range from window selections to photovoltaic panels. Each item you add into your callout list gives you and the any viewer the ability to run through materials and their specifications needed for the design process. By providing a list, the designer can easily understand each part added on to the building and give insight into what the building is composed of. Each callout or "take off" from either the use of a section or elevation, shows in detail the image of the material and the functional use. When creating a callout sheet, take a trace over one of your sections or elevations and place small circles on items incorporated into the drawing. Then extend two lines out of the circle from either side and create a larger side view drawing that may be either a circle or a square. This gives you the ability to create your own callout function and shows the materiality in the building. Each bubble or zoomed image can be drawn into a visual graphic of what the item is, and next to those images, you may be able to provide notes detailing item specifications. Detail the drawing and make sure you are incorporating every bit of detail you can provide to illustrate what you are depicting.

Floor plans, elevations, structural sections, callouts, 3-D drawings, volumetric drawings, and perspectives have all been created. Now let's move on to the control and the organization of documents. With all freehand drawing documents at hand, we can introduce a way to efficiently and functionally organizing these documents to advance into the next chapter.

C) Documentation and Document Control

The following chapter introduces the methods and procedures for controlling document flow and filing tasks. This is a way to control and manage all documents that were previously produced, drawn, replicated, or copied. The document control method is a procurement method that would assist the designer throughout the process of associating working drawings and notes together to consolidate tasks. Throughout the current practice, filing resources have proven to become difficult to manage with all the inventory of working drawing sheets and files. The studio for that matter and task of securing documents are part of an office management duty that does become confusing. To mitigate the loss of hard-copy documents, appropriate filing methods are needed. The filing techniques that are used in common practice are linked with simple filling methods by using products right off the shelf. Filing cabinets, storage system shelving, storage bins, storage spaces, and filing containers all serve to keep documents under control and in proper condition. To reduce the loss of documents, several functions can be implemented safely and effectively. First, an alphabetical document control system can be used with tape features or markings on folders. This can function as a method to select which documents can be accessed quickly. Second, a table can be produced to organize the location of documents, their accumulated quantity in each folder, and their respective type. The type of filing for document control gives authority to where the documents are located and how they can be accessed. There are no rules to this, as it only pertains to how well you can produce what you already can perform without accommodation. Document control can be used in these processes with the hard-copy documents that you are developing. If the process is for both hard copy and digital copy, a more suitable program can be developed.

D) Trace Overlay Drawings and Office Heath Precautions

Further along into the post-documentation process, there comes a point where drawings must go through evaluative measures to enhance specific aspects for the purpose of transferring them onto the drafting board. This is a lengthy process that involves going back into documentation and finding all previous drawings, including floor plans, sections, elevations, site plans, roof plans, and 3-D images. The process is one in which all drawings are reproduced by trace overlaying the drawings and placing them into drafting drawing categories.

It starts by taking all the completed freehand drawing from your files and numbering them based on their selective drawing type category as indicated earlier. Then, stage the drawings on a large table or pin-up board and go through a review of what you have. This will allow you to see everything in stages and will give you the ability to understand how you have finalized your complete sets of drawings. The review is a way for you to adjust and respond to your drawing techniques. It is also a way in which you can observe the results of your ideas through clear judgement on how things fit together and how they would be seen on a larger picture. The drawings you pin up should indicate to you these criteria for the next steps ahead in the design presentation process:

1) Completion—Check to see if all the freehand drawings you produced are complete and contain all aspects of your initial conceptual design. Drawings should contain completeness from start to finish and should have all parts and components together.

2) Clarity—Check to see if your drawings are clear and understandable to read. This would involve the review pin-up. It would be useful to have another person review it with a different set of eyes.

3) Cohesiveness—Check to see if all drawings are compatible and relate to each other. This is where your detail skills will be necessary as you view the drawings in correspondence with the level of accuracy of line in the drawing. The check may also include going back to freehand sketches to see if ideas were applied onto the freehand drawing or not.

4) Consistency—Check to see if everything in your drawings is balanced and has a consistent flow of line, scale, proportion, size, volume, detail, and geometry.

5. Revision—Revise any or all freehand drawings. When checking for mistakes or errors, it is good to go back through your sketch documents and notes. A designer tip to give out is that when something is missing, the best course of action is to go down the list. This makes awareness visible to the designer on how to revise individual parts of the full-scale drawings.

6) Review—The review process is ongoing, and it must be made on a case-by-case basis to understand the pattern of the design process. As an author's note, in design courses in architectural school, there are critiques that evidently show the cause and effect of design actions. In a review process, the critique methods can be implemented further by the pin-up of freehand drawings to gain more influence on how the drawings are interpreted.

7) Appeal—This is now the part where your work must be appealing to the eye. When you have gone through the steps toward making a direct connection between your drawings and the reader, the appeal factor should be made aware to the reader through how well the visible quality has been driven forward.

By attuning oneself to the requirements, one can perceive the value of how to develop and work toward a finite result. These are simply the parts of the practice that are both unique and commemorative to how drawings are produced in that they are reflective of the profession. For this, the reason behind making several steps in the process is that the outline of the practice is usually taught in schools by professors who know the design phases well. Lastly, for the sake of being self-taught, some of you wish to gain these skills as an accommodation to either your studies or your own practice. What you learn for yourself can be transmitted and signified to others without recourse.

Let me end by mentioning the word *phase*. Take into consideration your approach to your talent. Once you have done that, the best steps are ahead of you, and there are no phases that you will encounter if you do not lose sight of the very enjoyable talent that you possess. The word *phase* can bring a multitude of responses together, and it may as well be a characteristic profile of how you feel about the profession, which is very competitive and strenuous. It certainly involves many combining factors that illustrate the idea of a picture place in mind. As an author's note, I have gone through phases myself during my schooling where many days have been counted on the concentration of assignments and fluidity of activity from the brain. It is important to know that this is a field where the brain takes in a lot of information and sometimes all at once. So, I will provide some key precaution steps to take when being overwhelmed with the workload or study.

Architectural Engineering Mechanics

1) Plan your time. It is effective when you can plan your time on a weekly, daily, or monthly basis to include for yourself an allotted time where you can add as much detail as possible to the process that you are developing.

2) Take breaks. It is necessary to take intermittent pauses during your design process to self-reflect and to acclimate yourself to different nuances in the design process while being out of the picture.

3) Lose the sense of frustration. It can be very overwhelming at times to have to put things in place, organize, and sift through all the clutter. To prevent yourself from getting frustrated, it would be suitable to change the mood while working and after working. You can listen to music and be selective about what you listen to. Talk with yourself and get used to the fact that you are in the fast lane in this process. Nothing should be out of the picture. If you do get frustrated, there is always the silence of your mind to keep you at ease.

4) Exercise. Exercising helps relieve stress and helps reduce prolonged sitting anxieties. Getting up and stretching or pumping up some blood in the legs and arms helps prevent headaches.

5) Use a stress ball. The stress ball is a great little toy when your hands become stiff from excessive drawing, and it also can be used as a paperweight.

6) Pursue outdoor activities and brain stimulation. Getting out is important when you are stuck in the office for consecutive periods of time. Getting some fresh air, walking, and enjoying a drink or snack outdoors gets the motors running again.

7) Meditate. In the office, meditation time could be used to take pauses from work and to gauge the next steps ahead in the process. The meditation time also relaxes the mind and loosens up muscles that are constantly kept in position during work. Meditation is also proven to be helpful to relieve overwork loads by allowing one to focus on empty space or silent space while playing music or not.

8) Avoid the "locked up" feeling. Being an office junkie is a hard job itself, and over periods of time, it can have effects on the body, mind, and soul from being constantly at attention and in a particular posture. One thing to do is to sit up right when working to prevent body mishaps. While sitting upright, it is good to move the legs and calves up and down to increase blood flow. Bringing the shoulders back, placing the hands right above the kidney area, and applying pressure is helpful to reduce stress and strain.

9) Build endurance. Endurance is developed over long periods of time, where the focus and the commitment become second nature, and you may be saying that you need to hold on to the prevailing moods that surround you even if they are bad.

10) Reward yourself. Toward the end of your work, you will feel rewarded in a way. You might think that you have had the sense of accomplishment all along, and you might think that you have earned your reward through all your efforts. But one thing that is consistent with the reward or end accomplishment is that you yourself have made it possible beyond any discouragements that you may have faced. In the end, your results are a result of your consistency with your talent and no one else's!

DESIGN STRATEGY 3: PROGRAM DEVELOPMENT AND PRACTICE

Now that we have gone through the conceptual layout phases after most of your working drawings have been done, we can move onto the program development phase. This is a stage where the deciding factors of what is involved in your building or residence must be addressed. The program development phase is unique to the project in that the program, or what the project entails as far as its purpose, use of space, and function, can be developed into a set of identifying objectives used to advance into the drafting stages.

Architects and architectural engineers alike have used programs to govern their project details by showing legitimate concern for how spaces are developed and how a list of project rooms is specified for tailoring the architectural project together. The program is the incentive for starting any project and finishing it from its conceptual phases into its tiers of development. The program of a project consists of many variables that are all inclusive to the way a project is handled in its design phases. It is a way of giving the designer the ability to see ahead of time the items that are important to use toward the exponential guarantee of a project. For that matter, the purpose of the program is to deliver certain tangible requirements; these are those that are specific to the overall scope of the total project itself.

A program is used by architects to list the functionality and the constraints, which are the things that happen or are limited based on the main principle of the idea of the project. The entirety of a project must go through several junctions where the ideas of the concept are bound by the conditions of the site and the conditions of necessity for the project. The standard to follow through a program would be to simply identify the important objectives and how things got to belong to each other and how these things are compatible with the general degree of composition. The understanding of this composition is to fulfill the needs of the program and deliver the items needed for developing the building or residence.

A) Architectural Program Development: Designing

The program consists of many categorical items related to the building and its composition. When determining what the program needs, the designer should base his or her decision-making process on the client's or provider's requests. The program again can be in document form and can also be a set list of requirements that include the proposition of a built item. Here is an example of a preliminary program. This may also be indicated as a set of parameters for how much is involved in the building. This program is made by paying attention to the items that require need or suitability toward the project:

Program of Architectural Requirements

Objective: Design of an addition to an existing art gallery in New York City. The addition would be an extension of an existing adjacent gallery to the west end of the building. This addition is a complementary structure that would accommodate more gallery space by including several floors containing large open gallery space for various types of artwork.

Location: A famous museum located on the east side of Manhattan, which is also directly adjacent to Central Park and is settled in the midtown area.

Directions: 1071 Fifth Avenue, New York, New York, 1981

Program Deliverables:

Site—The site includes an already existing museum that is surrounded by nineteenth-century modern-era buildings. The museum is an abstract composition of forms and a winding tower that is cylindrical and is cut through its layered masses. The museum is the Solomon R. Guggenheim Art Museum designed by Frank Lloyd Wright and Partners. The addition is a collective involvement of space and form that would be accommodative to the museum and would reflect the architectural attributes of the building by enhancing dimension, form, and composition. The addition is by Gwathmey Siegel and Partners. The site is rectilinear, and is located on Fifth Avenue from 1071 Manhattan. It has a great view of Central Park and is the first building that stands out on street level by the apparent modern geometric architecture.

Existing Building: The existing famous Guggenheim Museum building is a complex arrangement of rectangular and upright forms. The main form that is unique to the building is the rotunda, and this is where, internally, a looping ramp is formed as a circulation system for people to walk on. Artwork would be presented on its interior walls and throughout other galleries. The museum in geometry is composed of one large rotunda maximizing the gallery space as a pinwheel-style building. The adjacent spaces to the west side of the building serve as lobbies, offices, restrooms, and smaller galleries. A main atrium space at the center of the rotunda provides a hollow, open area, and at the top, there is a clerestory window system designed by the architect himself.

Addition: Also noted as the Monitor building, this building was to serve and accommodate administrative, library, and other related functions in the building. It is prescribed as a large, rectangular structure that is upright and that allows for access to and from the existing museum with the use of ramps. Each floor of the Monitor building was established as exhibition-gallery space by integrating the pavilions with the large rotunda (Gwathmey Siegel 1982–1992, Rizzolli, New York).

Building Addition Criteria:

- nine total new levels on building
- each level serving distinct gallery spaces
- gallery spaces unique to the centrally driven rotunda
- each level with elevator entrance and ramp entrance
- floor-to-ceiling height levels
- interior lighting
- interior railings
- paneling
- structural columns interior
- water closets (restrooms)
- lobby area
- elevators and elevator area
- office space
- exterior *fenestration* on levels six, seven, and eight, this being a clear view of the exterior environment using paneled windows.

An articulated surface of the facade of the building is usually termed *fenestration*; it implies the particular treatment of the surface of the exterior walls.

Building Footprint Size

A maximum allowable size of XXXX square feet of building space. Total building footprint on site is XXXX feet by XXXX feet and total height of XXXX (dimension sizes not included for the purpose of showing example). The building footprint is a direct extension from the existing museum that is an ensemble of singularity, matching the site components of the main building.

Maximum Allowable Occupancy: The maximum allowable occupancy indicates the top tier of how many people are suitably able to fit inside the space. For the addition, an XXXX amount would be indicated. This amount varies depending on how many people occupy the building at any given time and at a percentage of circulation in the building.

Total Building Cost: Part of the program is the budget but sometimes not necessarily, depending on how the architect has taken into consideration all effects of the design and having considered relative cost versus design options. This goes further into detail in construction financial accounting and how architects establish their quantifiable price measures for their work performed. It is a little dicey and does require the extent of other practices involved. But for the consideration of the program cost, each item may be tailored to a bottom-line cost per design. This cost per design is up to you to register and research upon your own discernment as if you were to be in the position of design control. For the sake of architects or registered architects, the prevailing model of trend prices, design prices, and cost per item or material is something you have to delve into further. However, we can end with a total cost of items used in the process or each item at the given cost.

Zoning: The zoning is distinct in that special permission or authority must be given to the design professional to operate in the demanded area. The site contains a zoning order or a zoning footprint that gives the municipal authorities an idea of the type of site it is. As explained in earlier chapters, the zoning requirement must follow the city zoning board and the area must have been designated as a type of zoning area. This zone is a XXXX zone and is suitable for buildings that make up residential, office, or, in this case, mixed use.

Architectural Program—Line-by-Line-Item Document

System	Intermediate Factor	Operation	Cost	Units	Deliverable
Client	Guggenheim Museum	1071 5th ave New York, NY Manhatten	Est. $3.5 million	Sq. ft. XXXX	Addtition
Addition Design 1	Initial Designs	Initial Design and Conceptual Design	XXXX	n/a	Prelimanary Drawings
Addition Design 1	Preliminary Designs	Working Drawings and Drafting	XXXX	n/a	Working Drawings
Design Review	With Staff	Design Review-associate review	XXXX	n/a	Review notes+Drawings
Site	Site Review	Site plan evaluation	XXXX	XXXX	Drawings
Site Revised	Site Plan	Site plan development	XXXX	XXXX	Drawings
Gallery 1 Level 1	Floor Plan 1	Design of plan	XXXX	XXXX	Drawings
Gallery 2 Level 2	Floor Plan 2	Design of plan	XXXX	XXXX	Drawings
Gallery 3 Level 3	Floor Plan 3	Design of plan	XXXX	XXXX	Drawings
Gallery 4 Level 4	Floor Plan 4	Design of plan	XXXX	XXXX	Drawings
Gallery 5 Level 5	Floor Plan 5	Design of plan	XXXX	XXXX	Drawings
Gallery 6 Level 6	Floor Plan 6	Design of plan	XXXX	XXXX	Drawings
Gallery 7 Level 7	Floor Plan 7	Design of plan	XXXX	XXXX	Drawings
Gallery 8 Level 8	Floor Plan 8	Design of plan	XXXX	XXXX	Drawings
Upper Gal Level 9	Floor Plan 9	Design of plan	XXXX	XXXX	Drawings
Elevators	Design incorporation	applied design	XXXX	XXXX	Drawings
Restrooms	Design inc.	applied design	XXXX	XXXX	Drawings
Electrical	Design inc.	applied design	XXXX	XXXX	Drawings
Lighting	Design inc.	applied design	XXXX	XXXX	Drawings
Glass	Design inc.	applied design	XXXX	XXXX	Drawings
Doors	Design inc.	applied design	XXXX	XXXX	Drawings
Maximum Occupancy	Calc. Occupancy Design	Maximum Calcluated XXXXX occupancy	XXXX	XXXX	Drawings
HVAC	HVAC Design	HVAC applied Drawings	XXXX	XXXX	Drawings

The line-by-line method of developing a program may be more consistent with the objective work production of the design of the building. Using Microsoft Excel, you can produce such sheets by including the preceding items as seen in the image. The program is unique to the project of the addition and should cover all aspects of the building's design phase in combination with architectural building components and cost. This gives us a good look at the way a building program is developed and could be developed further into more tangible means.

The criteria mentioned in bold for the program are **system**, which calls for the type of building function; **intermediate factor**, which indicates what is expected from each system; **operation**, which indicates the work done for that specific line item; **cost**, which should be in a numerical value for the cost it took for each item from design to review to calculations; **units**, which can be measured in square feet or overall area; and **deliverables**, which should indicate the final product of the program outcome.

The main premise of the line-by-line program is to show the reader and the reviewer the subject terms of work being done by the designer. This is a unique design format that can offer better insights into the numerical and categorical identification of items for design. Not only does this indicate the essentials in the productive aspects of developing knowledge criteria, but it also looks nice! It is something to work off and something that can be revised and broken down again. Now the trick is to be as accurate as possible with your annotations and allow for the directive of the criteria to spell out what it is you wish to include in your overall design. You may wish to add other header criteria onto the table to show how each item has or will progress, considering the chapters included. But keep the bare minimum of requirements as indicated here. You may also wish to change "intermediate factors" to "performed function" or "applied item," whichever you prefer. The key strength of this program is in its readability in that it is clear to see what is being produced by the inclusion of item to end deliverable. This helps the designer with both the business end of things and the consistency of design.

B) Integration of Research and Influence

Architecture is a *process*. It goes without saying that the decisions and design choices are innumerable. The practice as it has served over thousands of years has revolved around basic principles of human form and material form. From ancient history to medieval times, the practice of architecture has held great merits to the external design applications of cultural antiquities. This is to say that that the living space and the external facade of buildings have been in transition from design to actuality all along centuries of change. Along with what is produced to provide some form of meaning to people, there are dilemmas that extend into every part of society. This is discussed here to note that the differences versus commonalities in architecture are so strong that they have rooted cultural impacts on those who live and preserve them.

Depending on how you grew up, when and what generation, architecture has been a balancing act between the old age and the new age, as it constantly serves as a reminder of the portrayal of progress and its symptoms. There are symptoms to progress that are embedded in the history of what was and into the history of what is or the current model. If you take for instance the city of Marseilles in France or any other major or minor geographical city in Europe, you will see an extended version of similarity in terms of regional and cultural style, tradition, color, form, and variance. The similarities are kept, providing a consistent character to the location, one that governs the attitude and the presence of the person inhabiting it.

In this chapter, architecture will be explored as it is a research method to understand and categorize these differences of person and place. How do these differences bring about further change for the person living, and how do these differences cause the need for change? In integration of research and influence of design, several examples will be discussed to highlight the importance of how these two are pertinent throughout the development of your presentations and for your own background knowledge.

1. **Research**

When we define the term *research*, there must be an emphasis on the definition of the word, which is to find and study new knowledge on any material in an effort to gain information about those subjects. Research is primarily a method that places the person in an activity that stimulates a gainful absorption of information through the direct means of study, reading, analyzing, evaluating, and defining what the subject is. The process is analogous to what you are doing now reading this sourcebook. It is for your own advancement and for the possibilities of reintegrating systems into your knowledge capacity by linking environmental concern and practical human observation together. In a framework of developing intelligence, research provides the way to accurately compose questions that are definitively uncertain to you by examining the outcomes and conditions of the research material.

In addition to the mental questionnaire process that you wish to satisfy, doing research is a step-ahead system that you can create to allow you to explore more into a specific subject or interrelated organization of subjects. It is hard to place everything all on one plate, and it may be condescending not to realize this being an architectural engineer whose overall responsibility is to consider all parts equal. In another words, architects and engineers, for this matter, rely on research to competitively advance in their own efforts. Their visions and tangible goals line up with what already exists, which in philosophy may be determined by the idea of the precedent. The precedent is termed here with the word "the" before it to include a rundown of architectural engineering as a highly compatible career field. The precedent is a notion of something existing before or something of a certain arrangement existing and having qualities of time, order, space, and condition.

When architects define the precedent, they determine objectively the correspondence to previous arrangements, and these arrangements may be in the form of existing buildings, traditional education in the fields, and compositions made by predecessors who are in their own time and history part of the lineage of a precedent. A precedent as an example can be used to identify a prior model of an idea, such as an obelisk, which is a long, solid, standing rectangular four-sided column with a pointed top. The obelisk can be understood as a structure and monument, which could be seen from its original inception during the polytheistic Egyptian era to its translation into nineteenth-century modernism at the Washington DC monument. As an example, the translation of an early or beginning idea to a later recreated idea brings the recognition of what the precedent reminds us of being. A precedent in architecture is the prior form or significance of interest that plays its part into more current forms of architecture.

In the process of research and with considerations to precedents, architecture is used to signify similarities in person, place, and purpose. When designers are immersed in their studies of architecture, their keen abilities to research are stimulated by the response to the current conditions that they wish to alter. Introducing a dynamic set of research options into the components of gathering information, whether it be from a book, the internet, or a captured photo gives the eye the understanding of a reality-based attention to detail about architecture. Architecture is formally seen nowadays as both a proponent tradition in the design of buildings and places and a historical reference to the ancient times. In this statement, there should be a concise acknowledgment of the way architecture relates to history. From historical architecture and from architecture as a lineage to today's standing structures, there is a definitive quality of resemblance. This is primarily attributed to the abilities of researching how and in what aspects these resemblances create what is today. As a footnote, this section is to enhance the creative mental process of stimulus for the fields of design and architecture by providing the essentials toward a direct process to research. This goes with understanding the second hand from the first in that whatever is found, read, and thought over may overwhelmingly become an "influence." This will be discussed further in the next part of this section.

To start research on what it is you wish to integrate into your building designs and for the genuine purpose of your own taste, you should look no further than your own home—the home that you live in, your home, and whatever it is that you consider as a home or a place of dwelling. It starts where you have your sense of space, your sense of comfort, and your sense of freedom. If these are not there and there are conditions that are impeding your ability to understand the home, try to prioritize yourself for your own benefit. Your own benefit is something that is discussed here for you to realize that your aspirations, hopes, and sense of self value should be addressed. It's like playing a trumpet; First, you decide that you want to play. You then purchase a trumpet and take lessons. There is no argument to this. This is something you wish to accomplish, and your ability to sound a loud trumpet comes from your source of value.

Gather a pen or pencil and notebook. The research process is one where you must spend your time providing relevant and useful information to your own experience. This requires attention to your writing and to your observations. What you will find through your research will be important for your design process. Note what you find, anything that comes to mind. And remember to provide detail.

The research process is like playing a trumpet, in that you are continuing to learn and continuing to perfect the use of the instrument to your best abilities. Yes, the process is difficult, when there are many others who can certainly play better and who might have learned more quickly than usual. But to get back to the essentials, your choice to play is solely dependent on your own terms. When introducing the research process to your studies, you must make considerations of where you wish to develop your knowledge and how you can transform what you know into what can make both you and your valued counteractivity a success. The research process is multiplicative; it is something that should challenge your sense of the world and your sense of place. Bear in mind, that I am, as the author, giving you good readers the information that can encourage your compassion for what you produce. And I will say to those readers who have reached this far and who are reading and dissecting to great extent these paragraphs in order to try to argue with nothing but the overall purpose of this book, return the book and shove your head in a sewer drain. This is a book, this a compilation of techniques, skills, and research to give people and especially students the ability to overcome obstacles. As other authors note, during my education, unknown to me were the barriers that limited such potential to exemplify words and actions into application. So again, for the readers who have read through up until this point, knowing how some readers are with what they read, they should do whatever they want. The aim of connecting with those who are interested in perfection is not hard to link by, and the research process is what can benefit you over the long run during your course of study or work.

To begin a research process, start by collecting a notebook and a writing utensil. Let us outline in detail the term *research process* for you to grasp a better idea of how this can format your designs and the material that you have selected to research. It would be a documented and controlled method to find information needed for your overall stages in design presentation. The items that you research will be overseen by you. This will allow you to take snippets and collages of what you have found to add into any portion of your designs. Let's not forget that the next section will offer you the ability to source your interest in what you select because it would be an influence on the overall style and programmatic design objectives you wish to incorporate.

Take your pencil or pen and notebook and start jotting down some objectives or goals you wish to have for the sake of your visual representation of your designs. These can be goals related to your building's form and function, your design etiquette for how you wish to interpret what you find and how you wish to use selected images or line work to your own design and integrate what was done before in certain styles and schemes, or what you wish to produce. These objectives will give you the references for how to investigate the details of your building, and they will also govern the historical reference points you make on how your building will be composed. Some key terms to know with definitions are *arrangement*, which is the way you put together or associate your research findings in your design. What you are trying to find is the next question. Look into famous architects who produced work during the mid-1900s into the 2000s. Look into historical architecture from the ancient times into European settlement, and then start following a pattern of research into the lineage of old versus new.

The more you investigate the past, the better you see the future in this respect, and in highlighting an aspect of modernism, you must understand the avant-garde principles of architecture, which were derived from the French and their migration. More closely, the 2000s have brought into perspective the cross-dimensional aspects of old traditional design in terms of a replicative and synergistic modernism. In your notes, develop what you find and take notes extensively on subjects of building materiality or what the buildings or structures were made of. Look into how they were built from internet videos and books—for example, the Duomo in Italy was a cathedral that was architecturally composed of primary forms and geometries using stone exteriors, glass, decorative glass, wood frame posts and beam constriction, and ornamentation.

As a researcher, you would obviously need to look at the building from photos or in real life and understand the form work at play and how it responds to interior and exterior conditions. If you do happen to come across floor plans of any building, make sure to print them out and have them documented for your learning process. Floor plan studies are a great way to allow the designer to see how architecture is developed and how people use space, as their presence counts for an occupation of use. Your research should also give you insight into how architecture and engineering have worked together to understand the man-made conditions in which a building or residence becomes proprietary for use in that the building serves an objective to provide its purpose as well as its public or private use.

In architecture, one of the key principles of the role of the shelter or place of habitat is the idea of public versus private. This is kept in consideration for the architect's role as a contributor to society by determining which criteria is needed to create an object or place defined under the scope of architecture, which is specific to the needs of living and use.

The public frame shows how buildings are made and used for a populace and how the building relates to public use. Public use can be described as a facilitation of place for the purposes of education, such as classrooms or libraries. Also, public use has many different explorations, and through the research, you will find that there are buildings with purposes apart from a house or residence.

Here are many of the types of public use buildings and facilities: library, classroom, church, mosque, synagogue, warehouse, university campus, grocery, restaurant, coffee shop, mall, plaza, city or town square, cathedrals, multistory buildings, mixed-use buildings (that serve a variety of purposes), retail buildings, shops, supply stores and outlets, restrooms, commercial warehouses, stadiums, open fields with buildings, skyscrapers, dormitories, hotels, resorts, apartments, multistory apartments, car dealerships, autobody stores and shops, government buildings, and private businesses. The research you make on what you find can show you the variances of building types depending on geography and limited to the time it was built.

Going further, you may find websites and look into books from today's perspective that have buildings that have transpired into high-tech capacities, such as industrial buildings, laboratories, corporate offices, and utilitarian industrial buildings.

The notetaking process should include where these buildings were made, the architects or builders, and any information you can gather from sourcing the original intent of the building. Find a building that you have seen visually and look it up on the internet. Then delve further into its construction. Every building has been designed by an architect or architectural engineer with the cooperation of a constriction manager or general contractor. In your research, define for yourself the people involved in the process. Define the year it was built, and try to get an understanding of how the age of the building fits into its style and purpose. Then investigate the numbers, square footages, total materials used, and cost. Books on architecture may include many images of buildings but also show the programmatic outline of how the building was conceived from start to finish. The authors may be the architects and may be critics of architecture who wish to highlight the architecture for marketing purposes. The defined writing of the architecture is a unique collaboration of associates in the fields that are describing the construction process of their building along with certain aspects of cost, financial decisions made, and relative considerations for public or private use.

Residences are always easy to work with because they are simple and have many styles that reflect organization and theme. The American vernacular is a completely authoritative style that governs many of North America's colonial settlement themes in residences. The vernacular, as it is called, is a description of the characteristic facade traits that colonial and Dutch-oriented-style homes have in conjunction with early British settlement in the colonies. The early Americans who settled in the country had buildings and homes built from practical observations, and they utilized the simple wood post and beam frames and wooden exterior covers.

Lastly, when you have developed a set of documents from printed materials online and your research, make sure you include the date when you found them, and make sure that these architectural buildings or places have an interest to your eye. What you gather from online sources or from books should be familiar to you in that you should feel as if you already know what the material is, and it should come to you as an experience through the senses that you can identify with the book or source as a learning effort. But even more so, the reading and viewing process should take you into the *world* that the building was based on.

With your notes, gather as much information as you can, and create a separation on another page by drawing a line down the center. This will be a task-oriented page where you can place on one side of the page what you have found from your research and what you would like to incorporate from that research on the other side. Architects utilize research to confer their ideas to their partners or associates and develop criteria that are essential to characterizing and enhancing their ideas. Of course, these would be somewhat illustrative to show during the presentation process of your buildings, but these are important for your own dissertations as to how you think the building should be shown. And if working with a client, the research you do can be brought into conversation with that client to see if there is a connection with his or her traits and desires.

2. Influence

Now that the scope of research has been defined, we must search for the one who has brought you toward your influence in design. This is important to know because your intent to produce is linked with some predecessor's qualities and ideas of how you measure your ideas. This may be from having a father or mother who has been involved in architecture or art from the beginning or having a near relative who possessed the same talent that you hold, or for the sake of mutual interest, perhaps you have found through your research that architecture fits the bill for you. Without going too crazy into what you like and trying to outline all the different architects or designers of the times, which is bizarre in respect to the overall picture of similarities that has developed over the years, you must identify for yourself someone and his or her style befitting yours.

Take for instance Frank Gehry. Who is he? Obviously, he is viewed as a "starchitect," as it is termed, indicating some architect who is a notable star. But looking into the beginnings of his work, we can see remarkable interpretations of his developmental idea, which stemmed from redoing his private home into some grandiose piece of work that later extended into his artistic compositions, such as the Bilbao Museum of Arts. Gehry emphasized wave forms and steel and aluminum paneling to show form and grace in his buildings. As an author's note, I would consider this bullshit architecture, which you will find a lot of.

Architecture has been in transformation because of individuals who have been influenced by each other based on their principal occurrences of time, place, modernism, and character. As you source difference architects, some of whom were very popular during the mid-1900s, you will see traces of architectural trends. Modernism as it developed during the Industrial Revolution came into the picture with a man named Le Corbusier, Charles-Édouard Jeanneret. His influence was totalitarian in the start of what was to be streamlined modern architecture of the twentieth century. A Swiss-born French architect, Le Corbusier was at the pinnacle of architectural greatness when it hit into the midcentury with architecture as the highest form of career known to man. Postmodernism developed as a sequential outlook to describe the new form and character that arose from modernism. The theoretical practices involved with certain designers from the times became the compartmentalized and unified conditions of how modernism transformed into more utilitarian and conventional systems.

Architects of those times, during the 1960s to the 1980s, were mainly brought up by the interest of old tradition, which was becoming lost in translation to large manufacturing capacities. And to that, the dynamic of industrial modernism had transpired into artistic and reformed style architecture.

Putting aside architecture for bit, the premise of this section is to introduce you to your world, as it is a combination of similarities that have developed over a span of thousands of years. When you are in your own capacity to learn, you are choosing to learn from what was done before. Your influences are who and what you are aspiring to be similar to and how you derive the characteristics from that influence to influence your work. Again, as we put architecture aside, we must adhere to the notion that you are a part of a society that reflects on itself in respect to what was done before and by whom. The aim of this book is to ensure that you learn what was done and how before you advance into any other stages of your own self-development. Your inclinations may extend from a friend's idea or even a stupid conversation forcing you to come unwillingly to your own defense as a layman, if you are, with a defiant perspective. Your influence is based on your conditions and your own capacity to create yourself as distinct from another. Your background is the key to your competence; it is how you grew up and with whom you choose to associate. Later, many of you who are in the stages of career development will know that it is a highly competitive life. For that matter, for you to be profound in your work, with your presentation skills, you must take advantage of ad hominins or counterarguments to yourself to understand where your influence lies. If you have an influence that is a particular person you are copying or replicating off, which is usually the case when it comes to your own objectives, you must get to the bottom of who you are as opposed to what was there before you. Ad hominins that are used to denigrate or to superimpose the self are based on the effort to compare the influence of a person to another's sense of esteem.

Influences in the architectural world are made quite often, as it is a practice in which people raise their eyebrows when someone is uniquely identifying him- or herself as a distinct individual. Thus, there are disparities in the process where competition is made unequal, which is a good thing because to each his or her own. The competition can become off-putting when comparisons are highly discretionary and result in severe criticism. Using ad hominins during your conversations for a search of influence can bar the similarity between you and your counterpart. This is a life of similarity beyond imagination. As an author's note, to mitigate anger and frustration developing over the years between counterparts who have excelled and who have declined, there should be a consideration into the architecture of self, as I as the author have indicated in the first manifesto, I wrote called *Determinant Theories and Philosophies in Architecture*. It is book that defines the self in a way in which the architecture of self plays an important role in how to facilitate your life objectives.

With influences, whether they be architectural or familial, the best consideration to keep is that you are solely responsible for your own actions and choices. This carries further into your life if you progress financially or in your career, as you will see that your obtainable goals are bound by your limits, and those limits may already be in place based on your upbringing. To counter this, the ad hominin of "Well, if you didn't have a degree in the first place, you would have ended up just the same as the next person without one" can show the level of comparison between one who has excelled versus one who has not. The purpose of this section is to identify that you are in the realm of attuning your skills to your own capacity, regardless of any title or constraint. This gives you the allowance to move forward in your own contentment as a person, and if there is the condition allowable for you to identify yourself as a distinct partner in the process, which I as the author am doing in this sourcebook for that matter, then by all means, put yourself out there. Just remember that you do not need to agree with anything you read or hear unless you deem it valuable to yourself.

The description of you and your influence is based on only one factor, and that is your own name.

C) Engineering Applications into Design

The interchangeable steps for introducing technology and the methods in which technologies are utilized into design comes from engineering. *Engineering* is an overly broad term, as there are many categorical ways of delineating engineering considerations into design. When the term *engineering* is used in the framework of design from applications of engineering principles, including the fundamentals in mathematics, physics, statistics, and, of course, analysis. This goes to say that in the analysis of principles for engineering applications, a great concern must be made as to how things work together. This section brings into perspective the engineering methods needed to devise and create executable plans that are in the design stages and how to clearly identify items of importance relative to materiality, structure, design compatibility, design performance, design constructability, and most importantly, conventional relationships to real-world design. Acknowledging the trace of engineering as a pinpoint system that allows for compatible designs to be integrated into the primary concept is part of this picture. This is where the parts of a design may need further articulation into refined dimensional analyses of size and overall dimensions. Taking engineering into the design aspects of buildings occurs during the formation of plans and specifications.

As we will note, plans are drawn again in accordance with the architectural designer's knowledge and expertise of building systems, interior and exterior special relationships in plan, construction detail, and composite design interpreted on a 2-D format. The 2-D representations of images are compiled in a set of drawings as mentioned in earlier chapters to signify the delineation of all aspects of the building being produced. Buildings comprise many aspects of technology and function out of the direct implementation of engineering principles in design. We will discuss how plans, specifications, construction documentation, construction details, annotations, site conditions, and technology encompass the engineering application into design.

First off, we have outlined the major and minor steps to take in preparation for your designs and the overall architectural development of your presentations. Throughout this sourcebook, the indications of detail and awareness to directives are made for you to gather as a resource for your further replicative development. As for most of you who come from engineering backgrounds, who have taken courses, and who have participated in lectures and presentations, you will know that a strong backdrop of historical reference is always made first as the implicative objective of any further teachings. This means that in any form of academic learning, with it being the professional take into career objectives, the rooted and fundamental concepts of knowing the advantages of the fields come hand in hand with applying them to the real world. Theoretically, as described earlier in a way to stimulate your thinking process, the solutions you find in this book are based off human experience in the applications of the design world. This is to say simply that anyone can pick up a pencil and draw, but what is important to identify is the intent in what is being produced. It is difficult to constantly replace words with actions in this field, as both are dependent on each other and are continuously on edge for improvement. Being a highly competitive field, the aspects of engineering play a significant role in the overall production of work in that anything produced is articulated to relate to the dynamic principles of mathematics and laws of geometry.

Plans are made by architects who usually take a good amount of time to produce effective and lean drawings. These plans have definite outcomes in that they must meet the requirements of both a universal category of human utilization and a conventionality into operative uses that are designated by state, federal, or municipal laws. Plans are organized to indicate detail and clarity in what is being produced, and the challenge always occurs when considerations are not made actual and instead are deemed as a failure. The plans must follow engineering requirements that are essential to incorporating a justifiable sequence of understandable drawings. These then are interpreted by developers and contractors and are further critiqued to "tweak" or modify aspects of the design in lieu of external considerations. In this business, it's not easy as an architectural engineer to always do what you want as opposed to what the client wants. This goes with saying that you must place yourself in a category of obedience to the fields in way in which you are responsible for your ego as well as your wallet. Of course, we have variances in the fields and the progressive architects who excel financially in the commercial sectors can define for themselves their prospects of success without much constraint.

It goes to show how much of an impact the dollar has in this field, as change comes with a price, and to that change, the price goes to the bank. Reading this material for that matter "can" concern the implications of your goals being related to external premises, either those who are in the position to recognize your objective or those who impede your objective. Plans occur as an isolated facet of the product of design in that the integration of worldly knowledge is presented as a further resource. The plans of a building must be flexible to the needs of the client, the principles of design and engineering, and the reflection of cost. When determining, for instance, how much space is needed for the design of a museum, one must consider the collective resources that compile for a museum to be built and in effort must conform to the aspects of the real-world conditions that allow for a museum to be occupied with considerations.

The considerations are not usually termed as such, as the practice essentially aims to provide accommodation to users. This then goes into certain aspects of service-oriented architecture for users in which what is produced is feasible for the people occupying the space. Engineering is at the forefront of this endeavor, as it gives the balance to an actuality of systems used to further create the proponents of a business in building. Designating the use of the building in plans calls for a multitude of reasons why each part of the whole idea must fit together succinctly with the notion of its prolonged success. Detrimental to this is the door of failure and disproportionate concern for what cannot happen. Engineering and knowing what is compatible to the design is important to show.

Specifications are another way the objectives of the design are interpreted. The specs of the design of a building show the connections between materiality and installation type. The focus of where the development goes after the design of the plans and the criteria of detail needed to objectively show the finite aspects of plans in accordance with universal laws and measurements. The specifications give both the designer and the contractor an outlined look into the functionality of what is being presented to be actual. It is often where critical mistakes occur and where the placement of items is changed. Specifications give absolute definition in terms of the design projection and are linkable with the plans to show the engineering aspects' visibility to the reader for definite understanding. Engineering must come into the picture here to provide details on the large conception of the design and to relate to the builders a method of communication that connects objects with words. This translation is very challenging, and many professionals are attuned to how these connections must be made. First off, the plans must be in sync with the specification. All materials and categories of items used in the plan must be reflected in the explanatory sequence of terms in the specs. The specs provide insight into the constructive engineering aspects of the design interpreted with notation and exploration of terms. Having specifications tailored to the plans and having them in response to the conditions of what is being built is the objective of the architectural engineer and must be made exemplary to show definite relationships.

As an example of how specifications coincide with plans and specs, we can give an example about "marriage or relationships." No, this is not a touchy subject; the purpose of bringing this into the picture is to show how one thing falls into the other's place. If you had a girlfriend or boyfriend, and your relationship was casual, meaning you had frequent intimacies and developments with that person, you would see that the considerations for each other were multiplicative. In the sense of specifications, what is being noted or gained further on is based on the relationship between the plans and specs—or in the case of two partners, the conditions of ultimate dependence on each other—if produced. Stirring up this further, as the author, I will say that each consideration is a continuous challenge that stems from the initial conception of the idea of connection. With plans and specs, the level of detail must be made to clarify the objective of the progress of the work. The intimacy as a condition of play between the two partners is objective and must be made actual in terms of connection and having it work. Knowing the interdependence on both ends, one must fully consider the weight of issues bound by the linkable characteristics of detail needed to understand where words resolve into actions. The specs play an important role in the overall determination of which classifications need to be made based on the information of the plan.

A typical specification or design specification outlines how a building is built and is organized to show dimensions of components, measurements, and annotations derived from the plans. Specifications are written out usually by the architect in charge to detail the explanations of what belongs where and how each part or piece of the built structure is connected. Specs are then sent to contractors and engineers for verification and review of the design being conceived and to make determinations if the design requires changes or additional information. There are also performance specifications, where the descriptions of the use of the design are described and help explain the intent of the project. These specifications are usually in sync with the design specifications and can also be altered based on any changes made to the plan, additions in drawings, or annotations made on drawings to provide a level of accuracy and relation to the plans. Performance specifications usually entail diagrammatic and annotated writings that govern the application of the project in terms of its conventional use.

Engineering is essential throughout the development of design in your presentations, and by understanding the concepts through theses strategy chapters, you will in fact be able to identify for yourself the components that are necessary to allocate in a tangible way. The study material you have already experienced throughout your education will give you the understanding of how function and method work together to produce a project. Communication is, of course, the most essential in this endeavor, as it plays a consistent role in determining the detail of directions needed to complete objectives in design. In terms of integrating your education in response to the demands of what you wish to produce, look further than the courses you have taken in your program and extrapolate what you have learned. Allow it to provide notes as to how you wish to have those courses be a part of your design presentation. Education is where the fundamentals of your knowledge are produced, and it is where the governing factors of techniques are made relevant to the overall composition of your project.

For instance, if you are developing a strategy to use the maximum allowable square footage of your building, you must go through an engineering process on how spaces are designed in sync with each other and how each space is utilized in dimension as a total in the maximum given square footage you can build in. This would take several notetaking abilities and would be reinforced in the conceptual design phase, where you would have to sit down and study the use of each room and what each room may require as far as materials, accessories, or machinery. Developing the considerations for these spaces is also determined by regulating codes and standards of building codes accessible to the building and controlled by requirements. Codes must be compliant. That means that whatever is being built must follow set regulations in size, scheme, structure, and safety. As for your intent on what your project is supposed to be, you must research the building codes that will determine factors such as square footages, door and window sizes, structural assembly and safety precautions in structures, maximum allowable occupancy, and structural loading, which is a practice relating to structural engineering that will not be discussed in this sourcebook. Take the time to find books on building codes, and research federal regulations for codes enforced throughout your state or country.

International codes vary depending on what country you live in and the considerations for how buildings are built based on their unit of measurement, their government laws, and their enforceable safety precautions. OSHA and ANSI in the use are two safety and regulatory agencies that provide guidelines on aspects of products and services for the construction industry. They detail safety precautions for how people may work in risk-involved areas. OSHA safety is a national effort made to regulate safety practices and train persons involved in risky and hazard related work by incorporating testing, documentation and training to persons at work and controlling aspects of work safety in order for proper administration to occur as a result or likelihood of failures. Preliminary Presentation Techniques

The last section of this strategy chapter is to outline how early stages of presentation skills can be used into the next and final chapters. The process should have included the workable steps to develop unique and tailored working drawings with the inclusion of engineering principles. The early portion of your presentation should also be a plan, which you need to decide on as the designer, for the showcase method for your final presentation. Think of it like you are being a director and editor while you are in the process of cutting and adding scenes with the use of a pin board. Images, diagrams, and snapshots should all be considered in this process to further enable you to collage what you have developed into workable sets of drawings. The chapter following this is a complete optional addition to your presentations, as it discusses the fine-line aspect of creating construction drawings using pencil and overlining with pen. This is how many students excelled in early architectural schools with their hand drawings and was the method I, the author, had incorporated throughout my schooling. With the advent of computers and programs that translate line work into digital line, as mentioned in the first sections of the book, this will be the clincher piece to this sourcebook. The aim, again, is to allow you as the designer to develop working drawings at your desk and familiarize yourself with the tools and methods presented in this book to allow you to work side by side with your drawings and have them uploaded into the software.

Now, going back into this section, we will discuss how to organize your work and how to make yourself familiar with the requirements of presentation. As mentioned before, the chapter following is an optional avenue for you to take, and by that, it would indicate the hard-lined aspect of drafting skills needed to either showcase your work in an ink format or to be able to mess around with drafting as a hobby. Both are included; however, this book focuses on the function of including the drawings in computer software, mainly as will be discussed in the section on the Revit Architecture program by Autodesk.

The preliminary stages of your presentation must undergo some level of thought as to how you need to bring together the components that make the presentation a presentation. This starts off with making sure everything you have for your project has been filed and numbered for you to retrieve and review. Simplifying this further, your work would be the resource for you to take the next step into diagrammatically producing the conventional—or real world—phase of your design. This means that your line-making skills must be perfect and must have the highest level of accuracy and dependency on clarity as possible. To facilitate this idea, we must understand the *mind* of the architectural designer.

An architect or architectural designer is a very disobedient person by his- or herself in that the formalities of regulation and constraint make the designer wish he or she could spray fruit punch out of a chimney or have caricatures of dogs and elephants on the roof of a building. The architectural designer has the absurd freedom to guesstimate his or her abilities in nonsensible and sensible ways in that he or she is bound by the constraints of line and regulation. This is being said because architectural designers have a mediocrity to themselves as people in that think as they perform and perform when they are under conditions of thought that relate to people and occupation. What will be said here before the outline of presentation skills is that the idea of this sourcebook is to define practically the functions needed to simply produce work without animosity and without bias to laymen or other fields. It also is provided to deter the reader from the antiquated personas that architects build for themselves as being quintessential or non-characteristic. Some architects are startup perverts and do a phenomenal job at being those. You know who you are, and shame on you. Get a new major or new job because you probably have not tipped the waiter. They do a phenomenal job of being offenders of the practice, as they tend to modify their skills to deter the nonsensical aspects of the practice, and that goes to show that a sense of desperation is evident in their own imagination. This is a sourcebook to provide the rules for imagining and give you the stronghold on which to anchor your talent. Architectural engineering has many facets of freedom, and these must be identified here to show you that the practice is concrete and revolves around core engineering and not despairing slapstick innuendos that only serve to rectify a sense of architectural perversion. Yes, we said it here. Draw a tooth fairy well, and do not forget her clothes.

Now, as for how to start your initial review of material for your presentation, give some insight into how you wish to promote your ideas in a sensible way. Then get into the nitty-gritty aspects of the presentation, as it will be you who will be in the hot seat for what comes next. We start off here by indicating your level of proficiency in all that has been discussed earlier, and with that said, you should feel like you have a good idea on the background. I emphasize the term *background* to make you aware that you hold this proficiency as a result of your own input. Then, steer toward including your self-assertion throughout the preliminary ideas of your presentation abilities, which will govern your motives.

Your motives are important; these are where and how you wish to word your ideas throughout your presentation. Keep always in mind that you must hold a sense of composure, being yourself as a respectful and considerate thinker and listener. Then put yourself to the test and allow your knowledge to be the backdrop for your stability in conversing fluidly with respect to your own comfort level. Your comfort at doing this, as will be mentioned here, is not the priority; however, your stability in showing that you are comfortable and that you can later be confident is the goal. You must take into consideration your clarity and your ability to transcend words, elevating them from their minuscule meaning to their bigger-picture quality.

There will be those who desire to initiate conversation with you and explore how you think in terms of your design, the decisions you made, how you made them, and why. The best course of action is not to follow conversation, knowing what is expected to be visible, but to know that your own expectations have met the criteria of what a substantial design may be. Be confident in your approach, and be confident in your appeal to your audience, as you may notice that describing what you have drawn may be more complicated than it looks. In that case, you need to feel obliged to your work. Then you need to promote your talent as undefined and characteristic to who you are. Yes, the process is difficult, and the process is objective in that your achievement may or may not be taken into consideration at all by critiques, former students, professors, or design professionals. You must be confident in your presentation, and you must not allow yourself to be diminutively judged, as you have taken into consideration the considerations that were first there to be explored for yourself. Engage with your skills by knowing where your weaknesses are, and those may be in the form of your initial concept or may be from your inability to clarify yourself based on your drawings. This must be reinforced by your attitude toward how you perceive yourself in front of people who are just looking at the drawing and hearing from you what it is so unique about it.

The hard part is when you can't help yourself but explain in basic terms what it is you are trying to develop. Rather, indicate to the best of your abilities what is similar to you and should be of that kind. For instance, as a designer, if you run into conversations where you feel you are being embarrassed by someone who is critiquing you, look no further than your options in front of you. Those options are your clear voice, your stand, your representation of self, your sense of humor if developed or not, and your guard if your humor does not work or if you know how to make what is humorous exemplary.

The preliminary presentation methods described here will encompass how well you attain your credibility toward your design initiative. Keep a discipline toward what you aim to show and keep a discipline to how you keep yourself in your best shape to take on the challenges of critical analysis. If you are in architectural school and you're around your twenties, you need to realize that you need to keep to the discipline, which is architecture, and the profession, which is the most complex, the most detailed, and the most profound of all the majors. I say this because this is the field of challenges, and this is the field of fundamentals in decision making.

As far as you know for yourself, your life has been full of choices from when you first had the freedom to think for yourself. A late professor of mine presented this idea in my first class in architectural school. A mention about the choices a person makes was laid out in front of our class to help us identify where things went. My professor explained to us that choices are limitless and that in limitlessness, there is a limit to identify. After he gave us a printout with a full paragraph of words showing actions and tasks, the headline stated, "Every day, you make a choice." After reading this and having a clearer insight into it, as I did back then in 2006, I regurgitated every bit of that sentence throughout my life as a student and designer. I took the fragments of those words out of necessity and survival to prolong my education and to lengthen my abilities to complete tasks completely. I took advantage of that sentence, and it has brought me to the compilation of this sourcebook. I hope that in its entirety, it will become of use to those who wish to both entertain their wisdom and rectify their talents.

As I am now in my early thirties, I see that the real struggle is not in the level of design or conceptual thinking, and as one gets older as a designer and as an intellect, the abilities do so also start to diminish. That is why it is important to engage your mind and your body in creativity. It is where all ideas and progress begin. Creativity must be the first ideal rationality behind your ideas, and communication is by part just a tool to render that idea as you move forward. Communication is unbalanced, infrequent, lacing characteristics and personality, and you must change that in your preliminary presentations. You must be able to add flair to yourself and give yourself the ability to commandeer your wisdom to tackle issues that may be sensitive or that may be embarrassing.

You must increase the rigor of your mind with progress in order to make progressive language be effective as a tool for your progress. One can say, "I didn't think that this would be so hard until I looked at how much is involved just to make a drawing come to life." Instead, the better way of saying this would be, "Yes, it's hard to understand this, but I can come to terms with this easily." One can also say, "You didn't do a good job at describing where X fits into Y and how you said before that Y was made because you thought it had to be there because there was no other way of putting that there without X there." Instead of hearing that and accepting that, you must confidently describe the answer as "This Y was initially here because X needed to be part of it, and the best way to understand Y is to know that X is in conjunction with Y." Conceive a plan to utilize toward your presentation abilities, whether it be in verbal form or written out. The idea is to prepare you and develop your communication and organizational skills needed to fulfill a presentation that is direct and to the point.

Identify what will be incorporated in your presentation. You will learn that you must add your freehand drawings, parti diagrams, and conceptual diagrams to the final presentation in the next chapter, but you must also consider the translation of the hand drawings into the computer. This will give you the upper hand in the presentation and will allow you to have an up-to-par and concise presentation.

On a pin-up board or a wall, post your drawings and review them for accuracy and detail. This process gives you the insight into how you should format your presentation and whether you require any additional changes. Each drawing you select toward your presentation must have a descriptive story or outline as to why it was drawn and how it fits into the big picture of your presentation. Take some time to note out what it is about the drawing you want to present and what it is that you want to say. Again, keep in mind that you must formally present this material to an audience or review board for approval of design. Review boards or critiques differ in that one is tailored to real-world projects and the other is for scholarly presentations. In a review board, committee members and site planners, such as civil engineers and architects, review the presentation to their discernment and judge the work based on the following criteria:

- clarity of presentation
- clarity of idea
- organization of material

- communication skills of the presenter
- content
- design quality
- design strengths
- design weaknesses
- overall presentation
- appeal or imagery

These criteria and many more follow the requirements for judging a quality of work based on the presenter's tactile and verbal abilities.

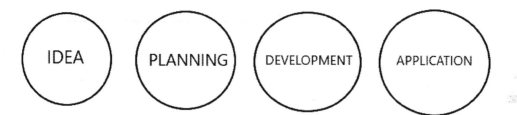

Four factors that will determine your prelimanary objectives are IDEA, PLANNING, DEVELOPMENT, and APPLICATION. Bring your objectives onto this focus and expand on each factor towards your success.

Your presentation skills are essential to how you formulate and execute your professional abilities for people who are listening to you. The audience will always pay attention to the way you first come off, and that is your first impression. Dress appropriately for the occasion, and make sure you add something that might catch their eye. Your presentation depends on your ability to concisely run through things from top to bottom and to equate what you know to what has already been done before. For example, you will be presenting as an architectural student or as a design professional, and you need to make your project match up with what you say.

Every word and every point of the finger to your designs should be planned. A great way to develop a presentation format would be to start off with cards. Memorize what it is you wish to say, from how you show your design concepts from the beginning of your project to how you developed your work consistently. Here are some guidelines on how to present topics in the right sequence.

Indicate the areas where you wish to converse.

1) Concept: Discuss the formation of your concept by bringing out what influenced the idea or the stems of ideas. Speak with authority and deliver the concept from its early life stage into language. The concept should clearly identify the representation for your

project, how it relates to the program you are acting on, how it is important in the discovery of what led you to create the ideas about your building or place, and what the concepts mean to the project in terms of identify and purpose.

2) Architecture: Define what is the architecture or what is the composition of what you have developed. This ties in with the concept and should register the idea of what you had with the idea of it being architecture. This should also be further communicated with influences or research as you should mention what it is that you have found through your studies that has led you to create such an idea. The architecture is your representation of what you have seen in your precedent works.

3) Precedent: Identify the precedent for your project, and if there are multiple precedents, put them together vividly through communication by clearly examining the ideas that were there before. Relate what you studied or what influenced you in your project by reinforcing aspects of those projects in your work through detail, design detail, bits and pieces of the work, and definite or humorous appeal.

4) Presentation: Go through your parti diagram and your concept and sketches first to show how you have created the idea. Then get into the nuts and bolts of the project, which would be the floor plans, sections, details, and any plan of the building. By purposefully examining each component, you are making the audience aware that you are the one who has designed it and the presentation criteria beforehand. It is better to start off with the floor plans to show where you have made spaces. Go through a quick sequence of how each space relates to each other and describe the important spaces of the floor plans to the best detail. Indicate how you made decisions about the plan size and elements of those plans in conjunction with the exterior conditions. Then if questions are asked further about those decisions, you will have room to explore them more with each drawing. Remember, the first presentation attempt should cover all or most aspects of the presentation and not be congested because the critics may wish to point out what you missed or are missing. Make sure you provide accurate information and clear language as to how you got from floor plans to sections to elevations and what prompted you to make those decisions. If it was the exterior of the building, indicate that you started from the parti and worked your way into the spaces. If it was the spaces, mention that you started with the plans and then worked into the volume or exterior representation of the building. Complete the presentation by going through each portion: site plan, floor plan, elevation, section, detail, construction structure, perspective, axonometric, and rendering.

5) Finalizing: Finish the presentation by restating the initial concept or idea, and make sure you include in your research summary what you have found to be an

influence on you. You should always make sure to not say "um" or wait too long during the parts of your presentation. Each part should be on a linear path of identifying the main ideas concisely and accurately. If you feel like you are getting butterflies in the beginning of the presentation, take a few deep breaths before, swallow, and hold firm in your presentation. Do not talk too loudly, but make sure you have some authority and confidence. You will experience gaps in your presentation skills that may become sensory to you as you try to formulate words into sentences. Sometimes you may lose track of where you are; in this case, review your cards and keep up with the linear aspect of your presentation being along a line of symmetry. In architecture, there are asymmetry and symmetry, which describe both a nonlinear progress (asymmetry) and a linear and organized progress (symmetry). Make an effort to keep things together and in a well-bound condition that allows you to maneuver from one topic to the other by stating the condition or issue and then providing detail afterward.

Proving yourself to someone takes a lot of work and involves challenging the judgment of others. Take your knowledge closer, step by step, into a sense of fulfilment toward what you wish to accomplish for yourself and realize the mental reward of your success.

Writing is a practice, and there is no true reason for why words must be kept in order *except* to "fulfill a sense of clarity and purpose." Writing is difficult, and throughout your life, you may have seen different models of the English language depending on which topic you are reading. To emphasize this further, the expression in writing follows the characteristics of the topic or study, and descriptions naturally become a part of a delineation that recognizes the language as being a constructive tool in the identification process of ideas. As the author, this side section conveniently describes the purpose of my writing style. I tend to write in the subjective frame of consciousness, where the thinking and doing happens. I do not create a facade of the language to manifest walls that cover details; rather, I show the details as a network of faculties that relate to the system of writing being a discovery method. This book is to allow you to think about yourself and your choices, and the writing I propose aims to put you in the hot seat of allowing you to apply your wisdom in both a practical and theoretical sense of things. The world of writing encompasses many deviations from the true components of which reading material ought to be composed. The language is structured in a way that balances an ebb and flow of sentences with periods and proceeding sentences with driven points.

I am merely expressing the writing through the subconscious mind by giving you the time to think externally from what is visible to you, rather than objectively view the writing as a sheltered system that diminishes any detail. In the writing here, I emphasize a corresponding balance of learning and utilization to further address the effort of the work involved to produce drawings and tangible documents. The writing is a way for me as the author to express both my interest in the field of design and the ability to translate designs into viable results. The problematic characteristics of writing tend to be from the lack of consistency and verbiage. By explaining in a coherent way, one can help others understand the fluidity of the writing. These chapters have been written to stimulate thinking, and they have been organized to provide background on what is being learned.

9

DESIGN STRATEGY 4: CONVENTIONAL DRAFTING TECHNIQUES

This chapter introduces what is essential or even *crucial* for both experimenting with and utilizing conventional drafting techniques in the design process. It is a chapter that provides the option for you to enjoy the opportunity to design your freehand drawings into well-produced sets of drafting sheets. Take the time now to consider what is important enough to your project to be shown visibly or on the two-dimensional canvas you are producing. There are exceptions to this in which your freehand drawings, ranging from site plans to elevation to floor plans may be interpreted on another larger piece of paper and detailed to a finer extent. But for the purpose of this chapter, a quick guide into the steps on how to prepare, assemble, and draft a drafting sheet will be explained. This is the first step toward a hand-drafted and finalized production of your work in which the ink will tell the story of your design.

Throughout the years, drafting has been a "basic determinant" for a method of producing fine line work with the assistance of drafting tools and equipment. The drafting table is the necessary tool for making this happen, as it has been used in offices around the world with the goal of producing various drawings to accommodate a compartmentalized project. The project is usually divided into distinct parts where draftsmen would spend their time sitting down and pricing line by line the controls for what their drawings are made of. During the Industrial Revolution, the draftsman was regarded as a member of an elite and uniform niche of people who represented architecture to the point. Draftsmen were trained in architectural schools and followed strict guidelines on the production of detail drawings and replications of sets of drawings. The draftsman honed his or her skills on the perfection of the line work. Drawing lines became a skill that involved accurate measurements of the instruments used and the precision to apply different line weights.

The use of the line is essential in the way drafting is done, and to know the right methods to produce proper line work, one must first understand how to incorporate line weight into drawings. By doing this, the designer can achieve the best results of his or her line work in a drafting drawing. The process involves the gradual use of the hand and wrist to provide pressure to the writing implement. One applied pressure or released pressure to apply light tonality. The technique is done to produce accurate line weights that are in sync with conventional methods of line work done in drafting. The line weight is measured completely by the hand-to-eye operation, which assumes that each line is different depending on the amount of pressure used to apply hardness to lightness. This can also be translated into darkness and lightness, as the pressure applied to the writing implement can be gradually tightened or loosened. This is important to know when drawing plans that require dark shades of line versus lighter shades of line, as it can play a factor in overlaying.

Take for instance overlaying HVAC systems onto a floor plan. You must first have the drawing of the floor plan ready, and then you must draw over the plan with a lighter line to show an overlap between the darkened line and the lighter line. When drawing HVAC systems or ceiling systems, it is good to coordinate these lines with lighter lines, as that can greatly change the scheme of the drawing into an overlay of systems. This is then a part of your set of drawings and can be used to visually show underlying or overlying systems in the drawing. The emphasis is to allow the hand to both apply hardness and subtleness into the drawing.

DRAFTING LINE WEIGHTS IN DRAWING

Hard   Light

Now that we have gone over a little history of drafting and the use of line weights, we will turn our attention to this chapter's contents as they relate to conventional drafting techniques. Toward studying architecture (at least for the sense of design and creating places of shelter or buildings, for that matter), this chapter is quick, painless, and straightforward. Each strategy subsection will go through the methods to which pencil-to-ink drawings can be produced on the full scale and with a drafting table. It is important to understand this as an option toward progressing into hand-drawn construction documents, which in turn may also be permissible for use toward the integration of today's architectural drafting software.

The fine-line pencil and ruler system of the drafting table is decades old, and most firms who participate in architectural design are few to be found with this method. Again, to reintegrate this to the purpose of the sourcebook, the following subsections are used to accommodate the designer for the purposes of conventional drafting. Take this into your designs and consider these options as a part of your presentation or as a part of including your freehand drawings into a more complete look. The considerations made into advancing your drawing abilities will go far, as you will encounter on the drafting table the varieties of choices needed to register complete sets of construction documents. As some of you may already be thinking, why go to such extents of drafting on the board again? My answer to you is to disregard this chapter or completely disregard your choice of reading the book in its entirety up to now.

As the author, I again wish to clarify the objective of the learning here and wish to aim for those who wish to learn and achieve the skills that can set them afoot into the more dynamic systems of digital composition. Feel free to think all you want, as you did purchase the book and you have taken your time to read what is apparent to the demonstrative needs of the student.

A) Pencil and Pen Trace Overlay

This section will quickly show how to demonstrate the trace overlay over your finalized drafting documents. Items here to consider are the following:

1) Freehand drawings you have produced to match the finalized drafting documents.
2) Large sheets of typically three by four feet for the pencil drafting
3) Large three-by-four-foot or varying size trace sheets
4) Light table or drafting table
5) Pencil, eraser, mechanical pencil, felt-tip pens in millimeter ranges for line weight

You can opt to use whatever size drafting sheet you like; however, for standard-sized sheets for construction documents, depending on the scale and size of your project, you might want to go for larger sheets.

The overlay process involves drawing to scale or to a larger scale the freehand drawings of your site plan, floor plans, elevation, sections, section details, detail drawings, roof plan, ceiling plan, mechanical plan, subfloor plan, section rendering, and all renderings in 3-D views. The overlay process starts by designating a larger scale for the drafting process and convenient sizes and dimensions into the pencil work for the draft.

Here is a sample of a simply drawn building to scale and with dimensions, sheet configurations, and annotations. The drafting drawing also includes other components to highlight the architectural drafting experience. One of them is freehand lettering, where capital letters and lowercase letters are written down through a guided line. The alphabet is presented to show some detail on lettering, and you can use whichever style you want to showcase your skills.

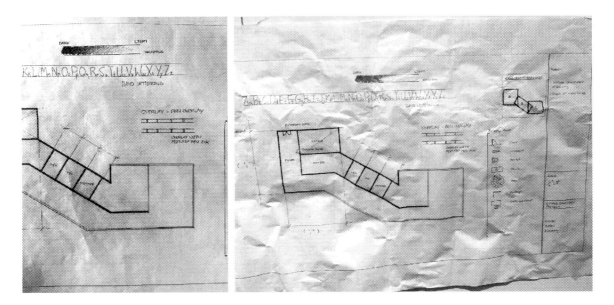

As you can see, there is a definition of a floor plan with corresponding parallel and vertical lines segmented off drawn walls showing dimension. There is also a legend, which is a great way to show the materiality and items of importance in the drawing. The document contains a side bar, which is typically used to show the company's title, architects, or engineers; the date created; and any information that is relevant to the date of completion or date of production. A scale is included to show the scale of the drafting drawing. Open spaces in the side bar can be used for annotative purposes, which can be numbered and pointed to locations on the plan given the type of plan produced. There is also a shade bar at the top, which shows the gradient of shade used for the line weight of the pencil.

B) Pencil and Pen Drafting

Now we will get into how to finalize one of these pencil drafts and turn it into ink. The objective is to produce black ink on the surface of the drafting sheet by using a trace overlay method that involves the larger piece of trace sheet. This is a three-step process, which would call for a lengthy draft, but it is convenient, and it is the best way to produce a clean, efficient, and inked large construction document. The process is lengthy and can be averted if there is the availability of graphite sheets that contain graphite, which can be traced over a drawing sheet. By this, the lightened graphite will be inked on the direct surface of the sheet. For the three-step process, start by converting your freehand drawings into size and drawing them onto the large trace sheet. Then once that is finished, a light table will be needed. Of course, if that is not accessible, your best bet is the graphite sheets or getting a large piece of glass. Be careful. Place a lamp underneath it, and trace over the trace with pencil. Once this is done, you can move onto the ink portion of the final sheet, which would be on a nontransparent sheet.

The process is straightforward. Move the protracting ruler (if there is one fitted onto the drafting table) left and right, and use a T square for the vertical lines. This will allow you to draft according to the right-angle position of the rulers. Architectural drafting is a practice in which everything must be in place and where everything must be set to an order of clarity, theme, and communication. Borders, north arrows, and floor plans must be drawn to a precise fine line with the use of the ruler and pencil or pen. The traditional draftsman relied on his accuracy of hand-to-eye coordination to fulfill the design objectives of his drawing, and by doing so, he was able to produce sets of complete and eloquent drawings.

Inking the stencil, as it is called, takes a considerable amount of time and patience. Wall thicknesses can be Poched (poh-she-d) in as they are produced with think black lines. We want to see those thick black lines on the sheet, so make sure you take your time to produce them. The trick is not to smudge the drawing. This can be done by cleaning your ruler every time a line is made with a napkin or preferably a dry fiber towel. Then produce your lines as much as you can with another napkin or small towel under your drawing hand, making sure that after your lines are drawn, you take some time to pause and let the ink set it.

This is the pencil and pen method, and it will create inked drafting drawings for your presentation.

C) Grid Use, Lettering, Symbols, and Detail Drawings

During the drafting process, it may also be easy to use the grid format for your layout of plans and designs. The grid is a square with crosshatched sets of vertical and horizontal lines that represent equal square spacings. The grid can be used in sync with your final documents and final presentation, which will also be explored in the last chapter. When using the *grid*, you may consider using it as a temporary tool with pencil or light pencil over your final presentation poster of boards. This would be a way for you to designate what goes where in your presentations. The following image is a conceptual graphic of how the use of a grid may help in designing spaces for where to put specific details of plans.

Square Grids and Designations

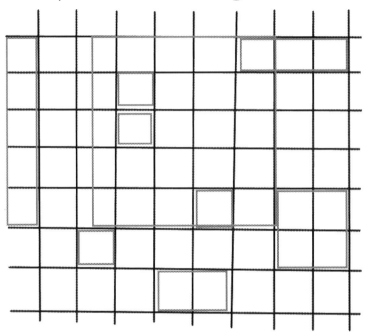

Lettering is often used to enhance a presentation through a hand-drawn graphic presentation of common letters and symbols. Lettering can be perfected through practice on plain white paper and with the help of guided parallel lines on top of each other. Try using these lines with a one-inch spacing in between them and write by hand in a casual style of lettering with the alphabet from A to Z, including lowercase. Cursive can also serve as a lettering agent, and perfection in the hand will bring out the best lettering.

Symbols are important to use in architecture, as they are representations of diagrammatic notes that are meant to provide key information about the conditions of the drawing. Symbols vary, and many of them are made to locate items of importance on the drafting drawing. They are small and are usually enhanced with the provision of a legend that indicates what the symbol is and what it does. Some symbols are digitally represented in digital prints as site conditions, plan symbols that reflect items on plan, and markers that indicate position and coordination.

Detail drawings are also important to add into the draft, as they are part of the whole of the presentation. Detail drawings are small snippets from the composition of the building and can be organized along the grid in distinct areas on the sheet. Usually, detail drawings follow some guidelines on materiality and information about the structural aspects of the plan.

These are some of the plan uses for architectural drafting, and they will be followed closer in the last chapter as to how to organize and execute these plans into presentation boards. The features that would be included from the architectural drafts would be organized on the boards in a way that would be presentable and effective to view.

D) Schedules

Some architectural presentations also include the use of a schedule that shows the time spent on producing and constructing the building or area. The schedule offers a variety of organized variables that can show the lengths of time made in the production of certain drawings, plans, and models. It can also be used to track the cost, flow of work, and amount of time it would take to build the architectural design. But for the sake of architectural presentations, schedules are designated as tools to effectively manage the time it took you to produce your drawings from concept to final presentation.

10

DESIGN STRATEGY 5: MODEL-MAKING AND THREE-DIMENSIONAL/ AXONOMETRIC DRAWINGS

This chapter will discuss yet another optional addition to your final presentation in the form of producing what is known as architectural model-making and three-dimensional drawings. This would focus on how to plan, create, and develop architectural models that would be representative of your plans. Architectural model-making is a highlight of the profession, and it takes constructive skills as well as a good mechanical adaptability to create models. Three-dimensional drawings are also considered as renderings or axonometrics can be developed as well to enhance the presentation. A three-dimensional drawing is a 3D version of your plan in which volume and sides as well as site conditions are explored in the We will explore both the model-making process and the three-dimensional drawing option for your presentation.

A) Model-Making Concepts

Model-making revolves around the idea of building or replicating what you have drawn in plan into an actual model. Models are generally made from materials that can be found at your nearest art supply store, and they are conveniently there for you to try your hand at. A model starts off with the plans you have produced, and to take this further, we must look at the floor plans, elevations, and sections to understand where the model would go as far as design. The floor plans are the best way to understand your composition, and by printing your plans and examining them in terms of design, you can make the transition into developing the model. This would take bringing the plans in and using them as a converted resource for your model.

Here are some materials used in the model-making process:

- mat board
- chip board

- clear plastic
- colored mat board
- perforated mat board
- mat board and design mat board
- superglue
- school glue
- scissors and cutting utensils
- cutting board
- light bulbs for light
- colored clear plastic
- straw or cylindrical wood pieces
- loose fill cardboard
- plastic tubes and plastic items
- cellophane
- wood
- colored paper

These items can be found at your art supply store, and they can be found at printing stores where scrap material is in excess. The materials are great for making models, and you will find that the versatility of the cut and the board can be made easier with the proper choices leading to their use. The objective is to produce accurate representations of your plans with what you have available to you.

B) Preliminary Three-Dimensional/Axonometric Drawings

Renderings and 3-D drawings can easily be produced using plans as mentioned earlier. To execute this, the designer must look into the dimensions and ranges of lengths of the plans as pertaining to wall lengths, heights, and widths and then generate either a one-point perspective or two-point perspective of the drawing. To do this, use a ruler to draw a straight line on a sheet at a given length, preferably long in length, and designate a point of the line at the far end of the line. For two-point perspectives, two points must be marked at each end equidistant from each other. Then designate the center line or a line vertical from the straight horizontal line you produce to create the height of your building. The line can be from the center for two-point or at the opposite end of the first point for one-point perspective. Then draw lines from the point to the tips of the vertical line from the marked line to show angles. From here, you can designate the heights of walls and other features of your building. This is a perspective drawing on a three-dimensional level. It is important to include scale in this in that the proper size must be determined by the size of your plans or converted.

Axonometric drawings are frame drawings of your building and can be drawn using the same two-point and single-point perspective as mentioned earlier. The graphic of the axonometric captures the essence of a three-dimensional object while at the same time being consistent with the idea of the use of line. Line work or lines are drawn in similarity to a perspective drawing; however, they show the detail of the framework or "skeleton" of the object. The axonometric drawing is a great visual tool to use to show structural aspects of your building model or to show some clarity on the spatial features of the composite form. Here are two examples of axonometric drawings, one in pencil and one in ink. These drawings show the similarities between the use of line and frame. Axonometric drawings are drawn using two-point or one-point perspective and then are carefully dimensioned to produce a framework of lines that show the internal components of your object or building.

It is helpful to designate what you are drawing specifically to which part of your building you are developing and then try to focus on that part as an axonometric. The drawing should be a graphical representation of a 3-D model of your building, which would consist of basic lines.

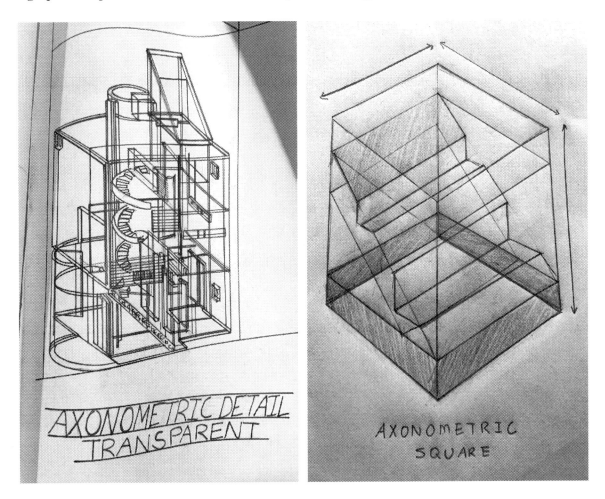

C) Model-Making Methods

Model-making is considered an art, and it is also a way to be creative in your presentation. There are a few rules to follow to develop models that can be either slightly like your designs and hopefully very similar. Model-making takes time and patience, but above all, it is something that requires skill and interest. When making a model, it should be a creative and engaging process, one you can play around with and experiment toward. A model is a representation of your design, and if it is a building, then you can interpret it as an extension of your design. If you are motivated enough to produce a model that fully resembles your plan ideas, then you must take into consideration the steps and the many resources you will need to fulfill this objective or goal or vision or plan. Whatever you want to call it, it's a model. Okay. You will spend a considerable amount of time planning and finding these resources to enhance your ability to produce models that are composed of basic resources. If you prefer to simplify your work, you can make models in different ways that will be explained in this section to allow you to show basic components of your design and advanced components.

Model-making is a process, and it is similar to converting plans into three-dimensional drawings. Again, if you are perturbed about this process, it is optional; however, it will be discussed here and made available to you as a designer.

The model-making process includes direct observations from plans with a sequence of drawings that can be converted into a physical model. It starts by gathering the floor plans, elevations, and sections and then noting their dimensions on their constructed heights, widths, and lengths of walls and exterior features. The interior features may be produced in several ways from the observations of the plans, and the interior form of the building can be modeled directly from what was drawn. Start by noting again the dimensions of the length, width, and height of the building from each elevation. Then print or create a copy of each floor plan to be laid out for the construction of interior spaces. This can be done by printing the floor plans to scale and then working on them by gluing and sizing the walls needed to fill each Poched (poh-sheid) wall from the plans. You will have to set each floor plan drawing separately in a prefabricated manner to produce each floor and then stack them on top of each other, having the plan then cut out of the model or removed.

Another way to produce each floor plan would be to designate each wall as a part or component of each cardboard or board used from dimensions gathered. This can be done by finding the sizes of each wall and cutting each wall to size and then overlaying them on a plan or a rough plan.

You want to consider making the model clean and free of any debris or excess glue. To do so, gather a cleaning item like a handkerchief or wet napkin to wipe away excess glue and to also clean up your area. This is pretty much a common chore for you, as you are the esteemed designer. Then organize your cardboard or pieces into groups on a table where you can effectively produce groups of parts that are needed to make a complete model. Gluing is essential in this process, so the more you take your time carefully measuring how much glue or adhesive to use—in the same manner of having attention to detail on drafting—the better your model will look.

Then start to be creative in your model by adding different elements from your building plans with different materials. This would enhance your model and can be a representation of certain materiality. Or if you are a realist, you can find components that resemble the real thing.

D) Model-Making Presentation

When a model is completed, it should have several components that would make it a real model. One is a solid base platform is necessary. The object or building should be firmly placed on top of it for presentation purposes and for the visual idea of ground level. Exterior environmental objects like trees, bushes, plants, people, cars, and so on can be added and found from art stores that supply model-making accessories. These would be added to provide more realism in the model and to attract the view of the reviewers in comparison with your plans. A model must also show either uniformity or color, and this is done by keeping all items the same material or making items of the model of different materially. The purpose of this comes down to the design presentation of the model and how you go about explaining your choices you made for the design of materials on the model.

Some models are uniform in material to show a conceptual base look that would be part of a scheme to represent development or form or mass. There are many types of models that can be produced in architecture, and most of them follow the same fabrication techniques. Development models are models that can be made one after the other and can show how conceptual ideas lead to final ideas. Massing models are uniform in construction and show the general shape or form of the structure you are building. Explorative models show ideas and concepts of your building through an exploration of materiality and notes. Presentation models are the final models to be produced, and they should show every level of detail from your plans.

Model-making is an art and can be perfected using techniques and proper cutting. Some models today are made from chipboard cut and processed through a laser-cutter machine. The laser cutter burns cuts in the chipboard from a digital or CAD graphic on software and then transfers it into the machine for cutting. This provides accurate measurements of pieces and perfect cutting. It would be preferred to go this route if you are creating a model that requires great detail and will be presented to a review board of professionals seeking to understand your building.

DESIGN STRATEGY 6: PLAN EXECUTION AND DIGITAL APPLICATION

11

This is the final design strategy chapter of this sourcebook. Now that you have made it this far in your reading and in your design, we will look at the advantage side of this sourcebook. This is the section where today's endeavor in design quality, design procedure, development, and skill are honed to specific applications. The digital divide, as it I call it, with respect to ego-critical professionals in the world of architecture, replaced hand drawing and conventional hand drafting with what is today's greatest tool, the computer. Thirty some years ago, when computers became the new facet of technological application, programs were created to replace, adjust, and conform to a digital production. CAD became popular as design software that created lines and mathematical geometries from parameters in the program that allowed for clear and objective qualities. Architects began using these software programs to design their buildings in accordance with the parameters of the software and were then able to develop their documents from these applications.

We have entered a time now where the digital commodity of computer software has completely replaced the traditional methods of design by hand. This is not to say at all that those methods are obsolete, and for the convenient purpose of reiterating the goal of this sourcebook, those activities can be in sync with the application of computer software. We will discuss the parameters and command features of AutoDesk Revit architecture, a software that enables direct modeling and design applications from commands close to Microsoft Office Suite commands. This is a program that allows for the selection of commands that are used to create walls, windows, doors, site conditions, constriction documents, and renderings.

Take a step back and understand that this is software that creates buildings with the simple selection of the command features. It also contains a variety of parameters that allow for adjusting dimensions, schemes, colors, renderings, and presentation format on digital sheets. It is a versatile tool and very accommodating to the designer as a system that operates on the ability to use tools that substitute lines for walls and lines as building items selected from the commands. I wish to be very specific in the writing here, as it will be defined in the several subsections as methods to understand and perform the use of the commands in conjunction with the hand drawings produced to accompany the design process.

Many designers today rely on straight dimensional analysis of their buildings or existing information from sites that can be noted or inventoried as information that can be input into the Revit software. This is common in the practice today, as it limits the amount of time and effort needed to make digital composites of buildings quickly and effectively. This, however, takes away from the working aspect of architectural design, and I will state clearly here that this sourcebook is compiled to develop a sequence of objectives leading to the translation of hand drawings into computer software. By doing this, the work is done, it is visible, and it is hourly or demonstrated to show an aspect of production. To clearly restate this again, because we have people in the profession who sneer at even the smallest of things, the sourcebook integrates what was drawn by hand into the direct application of the computer software. Now, as you have gone through your phases of design, you will come across this section as an ultimate registry of knowledge applied to a digital medium. This is to say that you will then be able to transfer what you have produced into the computer model.

A) Drawing Organization

We will go over how to manage and organize our drawings in an executive manner to designate what is being used in our final composition. The final composition is where we utilize the computer software made available to us as designers through what is again noted as the "digital divide" between traditional architectural formats and the more consistent platform of design delivery. The digital generation has swung into effect from the start of reputable companies who serve the same goals as we do as designers. Computer software that has excelled tremendously through its technical apparatuses has allowed achievements in the construction industry to soar to new levels. By this, the software has enabled the production of virtual models and real-life representations of plans and drawings to become tangibly genuine. One thing to know is that the theoretical assumptions that can be made as to how the evolution of these design tools developed can be understood by time and the placement of time.

From a philosophical point of view, with regard to the advantageous delivery of such technologies at our disposal, the software carries established place markers in a time line that has been reintroduced from previous software in the past. This is a mere explanation of how the discovery of new technologies has set new bars for the improvement aspect of developing mechanized and elaborate terms to utilize. We see, in this new decade, the upgradable features of both the smart phone and the desktop, and we do not fail to see how we have been led to such an environment, as it has been for the most part and with the coming of age of individuals representative of these technologies a place marker. Thinking philosophically about this even more, the attempt to reintroduce the software in new formats or in new collaborative parameters gives the end user the ultimate tool set for displaying the worth of their design proficiencies. In this new software, we see command features and pyramid parameters to be integrated in close attention to the unique development of the creation of the object.

The idea that has resonated in these software programs has provided an arcane facet of intellectual prowess for the user at hand, giving the sense of awareness of what the programs can offer not only to the dollar implemented toward the gainful requisition of progress made but to the display of empowerment that is at hand at the disposal of the user.

To discuss this further, organizing drawings is the first consideration to be made when attempting to manage both the software and the developmental path you are on from the readings offered in this sourcebook. This allows for the integration of accessible hand drawings to be consolidated and actively organized for the betterment of conversion and producing what is known as the digital composite. The digital composite has always been the development of such software, the computerized representation of the object. As described through architectural endeavors, the object represents a totality of placement, purpose, form, decision making, and actuality. This, in turn, is represented through the art of digital development, which can be further studied through many books published today as a means of discovering the advantageous tools required to competitively systematize what we learn into how we develop. This is to say that our knowledge is based on what we can develop for our tool systems to use in accordance with how these tools can benefit us, which is a strong statement that proposed a manifestation of certainties into the realms of imagination and creativity. It is also plausible to view this as a subjective rendition of how people can devise their work into manageable means of producing either the same thing or a reproduction of a model.

Now, to understand how to organize drawings, we must look at what was produced prior to integrating them into these software programs, and the software will be discussed in the next chapter with the attention of you the designer and reproduced with the permission of the companies to allow for the description of these software programs to be introduced in this book. This would further promote the feasibility of this book to be used as technical sourcebook in which methods are governed by traditional practices in what is being applied on the computer as I have said time and time again. It is important to note that the software being described is from previous endorsements made possible by the company and their textbook packages that offer insight on the use of the software revolving around architectural design. Not to digress too much into the objectives of the textbook or the companies' standards, but I am describing here the arrangement of these books to be useful in the essential utilization of conversion from hand drawing into digital work.

Organize your freehand drawings or full-scale ink and drafting drawings on a pin-up board or next to you on a large table. There is process here involving numbering and itemizing what you have produced to be part of the digital rendering process, which will be discussed in the next chapter. And for the record, the next chapter will go over the matching aspects of how these drawings can be associated with the software's features, namely the commands. Number each of your drawings according to importance and type. You may need to number sections of your drawings designated as sheet drawings for the purpose of using them as part of a compilation of floor plans and then sections, elevations, and so on. After numbering the drawings, you must look at what the drawings are and note that. Note them for what they are—for example, "floor plan 1," "east elevation," or "roof plan." Then effectively arrange the plans or layer your pages or sheets on top of each other for further use.

With your freehand drawings by your side as well, keep them for any minute details you wish to view when you are integrating them into the software. Now we are ready to make these lines straight, accurate, and real!

B) The Software by Autodesk

There is no information or material in this section that has been reproduced by any texts relating to Revit Architecture, and there is no unlawful reproduction of information derivative from any copyright material. This sourcebook is for commercial purposes, and the information presented here is informative, based on a preliminary and concise explanation of the introduced copyrighted command software from the Autodesk Revit Architecture program. I, as the author, have written this section to introduce the software without giving any details on the features, commands, or parameters or information about it. This is strictly for introduction to further use. Related reproduction of the material from these texts for commercial purposes is prohibited, and this disclosure is to emphasize the fair use rule under 17 US Code §106 for copyright holders. There is no reproduced information here. There is just an explanation of the product.

Autodesk has produced software that is compatible with the design building industry. This software offers a program that allows using design commands in place of actual building components. The premise of this section is to identify and explain without including cross-referenced material from the publisher to safely secure any written examination derived from their copyrighted materials. The software is available for students and professionals and comes at a price.

Autodesk Revit Architecture is used to design and plan models and building construction documents. This software enables the user to employ commands set in the software that produce design features. Without further explanation, I am completely aware of the implication being made in this writing to readers and professionals of the sort. The software is intended for designers to incorporate their designs and their expertise into the program by which construction documents may be produced.

You are in the position to further develop your work by using this software at your discretion. It demands accuracy, a steady control of the eye and hand, and a rigid acquiescence to the parameters of each command feature. By using the software, you are allowing yourself to learn and reproduce more accurate developments of your designs befitting commercial environments and procedures. The software is accessible to the designer and holds a multitude of operations that can be applied to a design construction document. It is at your disposal to make part of your presentation and your learning.

C) Drawing Transferring and Translation

It is a process, and once you have reached the potential to ultimately define your course of work, then you can proceed with the software. This is to identify the components necessary for translating and transferring information from your hand drawings into the software. By addressing both the translation and transferring aspect here, I am aiming to promote the sense of work involved, the developmental rationale needed to obtain those documents in order to have them side by side for reintroducing them into the software, making it compatible for the designer and for the effectiveness of the document.

The process starts by having your essential freehand documents gathered in place for you to access them and have them available to you to translate them into the software. This is done by gathering your design notes, your finite freehand drawings, and your architect's ruler. In doing so, you will be able to create the dimensions of your applied hand drawings in the software. Create a list or an extensive notebook of which items you wish to prioritize toward the application of the software's command features. I will not indicate the commands in this section, but I will organize the language to help you understand how to take your work and place it in the software.

You should have a collection of dimensions, floor plans, sections, and so on, and these must be translated into the software with the command features, walls to walls and windows to windows. Executing these commands requires the simple actions of identifying the components of the design and then introducing them into command features. Take a good amount of time to understand how this done. It takes a comparative analysis of plan and program to execute the line work, or in the case of Revit Architecture, the exact component of the design system.

The translation is made when you as the designer are sitting down and assessing your work as a part of your effort to continue to produce the scheme or idea that you wish to present. This is in part due to your abilities to coordinate the effective means of how you have hand drawn those designs and how you will take them to a digital composite. You want to be able to arrange your drawings in front of you or next to your computer, and with the aid of notes and a system, you can develop where itemized components are selected to be applied within the software. Pen marks to point out features and components can be translated onto the computer screen in case a reviewer is behind your shoulder. Then adjust your hand-drawn plans and proceed with applying them with the tap-based commands of the software.

This will give you the best ways to make your work visible, make your work full of content, and have your work be a part of your planned program and visual career model. In the office with other team members, you will be faced with decisions that you need to make quickly and without too much concern for the practice. This depends on your abilities to focus on materiality and take decisive action. When you are in the process of translating your work, you want to be mindful of others and their motives, as you will be in the hot seat of the command operation. Your team will contribute to your success by indicating where items are not visible and where they are visible. Teams may approach you directly and ask you how and why you are making the decisions you are toward the design. The theory behind this is that you are in control of the outcome, and the steps ahead of the outcome must be governed by the attitude of design and the intent of producing the composite resultant object.

D) Final Presentation

We have reached the point of no return. You have now assessed and asserted this sourcebook to be actual. This is an actual sourcebook defined under the conventional standards and practices of architectural engineering to the finite extents of utilizing each design method toward a complete and operable presentation. You will now find that after you have learned and used the software, given your proficiency and effort, you can rely on yourself to advance in any way, shape, or form the program needs to fulfill a design presentation involving architecture. This is an enjoyable process, first and foremost, and if you are not aware of this, then you have completely wasted your time reading this sourcebook. If you have read up until now to learn, then you are in a good position to complete your presentation. If you are the reader and you are merely concerned with the language, the position of detail, and the practice professionally explained in this sourcebook for your own corrective awareness, then pick up a new hobby.

The presentation includes all the required drawings, plans, and 3-D models, including hand-crafted models. You will need to finalize your work by indicating on your presentation sheets the size of sheet for the presentation, the proper sheet annotations and sheet requirements for constriction documentation, and your placement of design objects on the sheet.

Organize your sheets so that there is a maximum visibility to see your designs and to see everything about your work, finalized and fine-lined. If you have decided to present the material fully hand drawn, then again note that your line work must be accurate, and your designs must be fulfilling. If you have chosen to go into the computer program, then reiterate the translation and transferring components to optimize your design presentation.

This is the complete student design sourcebook for the practice of architectural engineering.

ARCHITECTURAL ENGINEERING THEORY: WORKING DRAWINGS AND A PERSONAL NOTE

12

Architectural engineering is a well-defined practice that has evolved over the course of many years into what is today's premise of architecture. Both architecture and engineering have been controversially examined under much academic awareness and compliance toward a constructive field that is specific to the change process of shelter. Thus, through education and study, the individual can be made aware, relating his or her experiences on the subject in part through the division of architectural principles and core engineering supplements. The design aspect revolves more into architecture and architectural language as the more concrete and mathematical components are relative to engineering.

There is much theory in architecture, as it explores many dimensions of person, place, and purpose. These are the three things that make up architecture, and they cannot be defined into words without any clear delineation of the objective interpretation of what they ought to be. Architecture has been examined and dissected into a subject matter that presents multifaceted connections with environments, money, and political concern. To understand architecture, you must first understand the methods of change, what change is, and what change does. Then you must interpret that change from the past into the future with your career objectives and your goals in mind. This has many implications that can appear in a variety of results. If you have money and you can produce thoughtful and meaningful architecture, then you have reached some aspect of a goal or you have reached your goal.

But if you are able to think and plan and make decisions to greatly impact a surrounding environment with the use of design and with the intent of gathering awareness toward what might be an artistic endeavor or a translation of an image from life, then you can be someone who has truly possessed the skills and work to be called an architectural engineer. The work involves years of dedication, listening to your intuition, and invoking your best senses of what place and identity are.

My personal note here is that I am still flummoxed by architecture and its overwhelming capacity to open the mind into space or spaces. Architecture is living in spaces and feeling good about what you see, feel, and wish to be around. Architecture is a component to the success-driven corporate world shadowed by money, materials, and profit. But there is architecture that is long lasting, a part of the land, aiding in people's activities and helping people to better their lives. I see architecture and have always seen it as an opportunity to be able to create and make things from learning from others and to be able to use the skills to build. I will not digress into ego, as it is the evil of this practice, but I will say that a good designer can see through anything.

Ata Asheghi

Printed in the United States
by Baker & Taylor Publisher Services